# Reservoir Control Operation and Water Resources Management

Guest Editors

**Yuxue Guo**

**Li Liu**

MDPI

Basel • Beijing • Wuhan • Barcelona • Belgrade • Novi Sad • Cluj • Manchester

*Guest Editors*

Yuxue Guo
College of Civil Engineering
and Architecture
Zjejiang Univversity
Hangzhou
China

Li Liu
College of Civil Engineering
and Architecture
Zhejiang Univversity
Hangzhou
China

*Editorial Office*
MDPI AG
Grosspeteranlage 5
4052 Basel, Switzerland

This is a reprint of the Special Issue, published open access by the journal *Water* (ISSN 2073-4441), freely accessible at: www.mdpi.com/journal/water/special_issues/ZEL5JEQZ46.

For citation purposes, cite each article independently as indicated on the article page online and as indicated below:

| Lastname, A.A.; Lastname, B.B. Article Title. *Journal Name* **Year**, *Volume Number*, Page Range. |
|---|

**ISBN 978-3-7258-4386-2 (Hbk)**
**ISBN 978-3-7258-4385-5 (PDF)**
**https://doi.org/10.3390/books978-3-7258-4385-5**

# Contents

# About the Editors

**Yuxue Guo**

Yuxue Guo, received her Ph.D. in Water Conservancy and Hydropower Engineering from Hohai University in 2019. She worked as a Full-Time Researcher and PhD Supervisor at Zhejiang University. She leads key projects including the NSFC Youth Program, National Key R&D Program Sub-project, and the Natural Science Foundation of Zhejiang Province. She has published over 30 papers in journals such as *Water Resources Research, Journal of Hydrology and Hydrology*, and *Earth System Sciences*. She holds over 10 patents and published 1 monograph. Her doctoral thesis received dual recognition as a National Outstanding Hydraulic Discipline Doctoral Dissertation.

**Li Liu**

Dr. Li Liu's current research interests include the forecasting of rainfall, river flow, and flood. The forecasting time-horizon spans from a few hours to several seasons. Her other research interests include distributed hydrological models, deep learning, and uncertainty quantification.

water

MDPI

*Editorial*

# Reservoir Control Operations and Water Resources Management

**Yuxue Guo and Li Liu ***

Institute of Water Science and Engineering, Civil Engineering, Zhejiang University, Hangzhou 310058, China; yuxueguo@zju.edu.cn
* Correspondence: li_liu@zju.edu.cn

## 1. Introduction

Water resources are among the most essential materials required for human survival and development. Water resources management is the activity of planning, developing, distributing, and managing the optimum use of water resources. Effective water resources management is crucial for sustaining ecosystems, supporting economic development, and ensuring social well-being [1]. It optimizes the use of water for food production and energy generation, ensuring that water, food, and energy systems have access to necessary resources while preserving environmental sustainability. Additionally, it strengthens resilience against natural disasters such as floods and droughts, promotes equitable access to resources, and upholds cultural and recreational values. Overall, it is essential for balancing human and environmental needs while ensuring long-term water availability [2]. However, water resources management is currently under increasing strain due to factors such as population growth, climate change impacts, and growing demand for water, food, and energy [3,4].

Reservoirs are a central component of water resources management, offering essential storage and regulation capabilities that support a range of water needs and environmental objectives. Optimal reservoir operation is one of the most effective non-structural measures for improving water utilization efficiency without requiring additional investment [5]. The successful operation of reservoirs and water resources requires a comprehensive understanding of modeling-related uncertainties and the integrative application of artificial intelligence technology to generate sustainable solutions for water, food, and energy systems in a changing environment [6,7]. Through the publication of this Special Issue, we seek to stimulate further research and dialogue on this critical topic, thereby contributing to global efforts aimed at achieving water security and sustainability.

## 2. Summary of the Contributions in This Special Issue

### *2.1. Reservoir Flood Control Operation*

Flood control operation is a significant way to prevent the threats from flood risks to ecology, environment, infrastructure, agriculture, and even human life. Three papers addressed new models in reservoir flood control operation (contributions 1–3). Contribution 1 developed a dual-objective five-reservoir operation model by considering the flooding risks both downstream of the basin and in the Miyun reservoir area. A parameterization–simulation–optimization approach was employed to obtain the Pareto-optimal front, providing a list of optimal rule parameters for customized risk preferences. They found that the current schemes result in a 10.5% higher upstream inundation loss and an unsatisfactory CNY 17 million of equivalent water transfer loss compared to the operating rule optimized in the study.

Contribution 2 presented a flood risk control method oriented towards floodwater utilization that considers multiple main flood risk factors. The proposed method not only achieved the specifications of the flood limited water level (FLWL) under various acceptable risks but also dynamically controlled the water level to enhance floodwater utilization.

They indicated that the proposed method raised the FLWL by 1.00 m above the present FLWL in the absence of flood risk. The available flood resources in both the wet and dry seasons increased, on average, by 0.83 and 0.81 billion $m^3$, and the flood risk remained within an acceptable range after raising the FLWL by 1.00 m.

In Contribution 3, a joint-optimized operation model of sluices in the group that combined "offline calculation" and "online search" was proposed to investigate the optimal sluice operation under high-intensity peak shaving and frequency regulation. The authors found that the total number of adjustments of the sluices of the cascade hydropower stations was reduced from 1195 to 675, a reduction of 43.5%, and the leading hydropower station, Pubugou, met water level control requirements, whereas the fluctuations in the water levels of the two downstream daily regulating hydropower stations, Shenxigou and Zhentouba, were reduced by 1.38 m and 0.55 m, respectively.

### 2.2. AI-Based New Monitoring Technology in Water Research

Climate change has exacerbated severe rainfall events, leading to rapid and unpredictable fluctuations in river water levels, necessitating the development of new monitoring technology (contributions 4 and 5). In Contribution 4, an innovative methodology was proposed that leveraged ResNet-50, a Convolutional Neural Network (CNN) model, to identify distinct water level features in Closed-Circuit Television (CCTV) river imagery under various weather conditions. They indicated that the method provided an accuracy range of 83.6% to 96%, with clear days providing the highest accuracy and heavy rainfall providing the lowest accuracy. The study introduced a promising real-time river water level monitoring solution, significantly contributing to flood control and disaster management strategies. Contribution 5 introduced a dam deformation prediction model based on a long short-term memory (LSTM) model with interferometric synthetic aperture radar (InSAR). With a case study in the Xiaolangdi reservoir, they revealed that the cumulative deformation accuracy was 95% compared with the on-site measurement data at the typical point P. The correlation between the reservoir level and dam deformation was found to be 0.81. The overall deformation error of the dam was predicted to be within 10 percent. The results showed that the combination of InSAR and LSTM could predict dam failure and prevent potential failure risks by adjusting the reservoir levels.

### 2.3. Changing Environmental Evaluation

Current global climate change and human activities have increased uncertainty in the hydrological cycle. In this changing environment, the risks and characteristics of flood, drought, and other disasters are varied, posing challenges to water safety (contributions 6–8). Contribution 6 investigated sedimentation characteristics under the influence of multiple factors in the main urban area of the Chongqing River section as a case study for the operation of cascade reservoirs in the Jinsha River. They found that the rate of sedimentation increased with sediment inflow, peak flow rate, and duration, while the location of sedimentation shifted as the concentration ratio changed. Contribution 7 analyzed the temporal and spatial evolution patterns of the baseflow through statistical analysis and the Mann–Kendall test. They found a higher baseflow contribution in upstream areas compared to downstream areas at both stations. The baseflow and BFI had significant upward trends in the dry season, while their trends were not uniform during the wet period. In Contribution 8, a new composite drought index was proposed that could comprehensively characterize meteorological and hydrological drought by combining the standardized precipitation index (SPI) and the standardized baseflow index (SBI). The results showed that the established composite drought index combined the advantages of SPI and SBI in drought forecasting, with an accuracy greater than 80% based on the trained random forest model. The study provided reliable and valid multivariate indicators for drought monitoring and could be applied to drought prediction in other regions.

### 2.4. Integrated Water Resources Management

Water resources management is a challenging task caused by huge uncertainties and complexities in hydrological processes and human activities (contributions 9 and 10). In Contribution 9, a multifaceted analytical framework comprising the CRITIC method, the standard deviation ellipse, the harmonious development coefficient, and the coupling coordination coefficient was developed to investigate spatiotemporal evolutionary trends and overarching harmonious development states between the high-quality economic development and water resource systems. They indicated that the epicenter of high-quality economic development indices was situated in the periphery of Lake Tai, whereas the fulcrum of the water resource system indices was in Huzhou City, both displaying a northwest–southeast orientation. Contribution 10 conducted a scientometric review and metasynthesis of the existing uncertainty analysis research for supporting hydrological modeling and water resources management through a co-citation, collaboration, and co-occurrence network study. They found that the USA contributed greatly to the publications and cooperated with most countries/territories. The Chinese Academy of Sciences was the leading institution and had a relatively intimate relationship with other institutions. The study also indicated that synthetical uncertainty management for hydrological models and water resource systems under climatic and land use change will continue to be studied.

## 3. Conclusions

This brief report provides a valuable overview of the ten selected papers featured in this Special Issue, with each making original contributions to advancing the state of the art in water resources management. The ten contributions encompass four major research subjects: (I) reservoir control operation, (II) artificial intelligence methods, (III) changing environmental evaluation, and (IV) integrated water resources management. They offer a high level of research and practical insights into the implementation of new methods and strategies for water resources management and reservoir operation, supported by case studies from various regions around the world. The future of research on water resources management hinges on a comprehensive understanding of environmental, socioeconomic, and technological factors. We hope that this Special Issue will encourage further research in this area, leading to improved methods and techniques and providing a deeper understanding of water resources management and reservoir operation.

**Author Contributions:** Conceptualization and organization, Y.G.; methodology, Y.G. and L.L. All authors have read and agreed to the published version of the manuscript.

**Data Availability Statement:** The data that support the findings of this study are available on request from the corresponding author upon reasonable request.

**Conflicts of Interest:** The author declares no conflicts of interest.

## List of Contributions

1. Wan, W.; Liu, Y.; Zheng, H.; Zhao, J.; Zhao, F.; Lu, Y. Optimization of Multi-Reservoir Flood Control Operating Rules: A Case Study for the Chaobai River Basin in China. *Water* **2023**, *15*, 2817.
2. Zhou, L.; Kang, L.; Hou, S.; Guo, J. Research on Flood Risk Control Methods and Reservoir Flood Control Operation Oriented Towards Floodwater Utilization. *Water* **2024**, *16*, 43.
3. Mou, S.; Qu, T.; Li, J.; Wen, X.; Liu, Y. Joint Optimal Use of Sluices of a Group of Cascade Hydropower Stations Under High-Intensity Peak Shaving and Frequency Regulation. *Water* **2024**, *16*, 275.
4. Chen, J.-F.; Liao, Y.-T.; Wang, P.-C. Development and Deployment of a Virtual Water Gauge System Utilizing the ResNet-50 Convolutional Neural Network for Real-Time River Water Level Monitoring: A Case Study of the Keelung River in Taiwan. *Water* **2024**, *16*, 158.

5. Fang, Z.; He, R.; Yu, H.; He, Z.; Pan, Y. Optimization of Reservoir Level Scheduling Based on InSAR-LSTM Deformation Prediction Model for Rockfill Dams. *Water* **2023**, *15*, 3384.
6. Li, J.; Ge, H.; Ping, Y.; Dong, X.; Zhu, L.; Ma, Y. Sedimentation Characteristics of the Fluctuating Backwater Area at the Tail of Cascade Reservoirs: A Case Study of the Three Gorges Reservoir. *Water* **2023**, *15*, 4011.
7. Xie, H.; Hu, H.; Xie, D.; Xu, B.; Chen, Y.; Zhou, Z.; Zhang, F.; Nie, H. Spatial and Temporal Assessment of Baseflow Based on Monthly Water Balance Modeling and Baseflow Separation. *Water* **2024**, *16*, 1437.
8. Huang, S.; Zhang, H.; Liu, Y.; Liu, W.; Wei, F.; Yang, C.; Ding, F.; Ye, J.; Nie, H.; Du, Y.; Chen, Y. Assessment of Hydrological and Meteorological Composite Drought Characteristics Based on Baseflow and Precipitation. *Water* **2024**, *16*, 1466.
9. Liu, D.; Dai, Q.; Yuan, G. Evolutionary Trends and Coordinated Development Analysis of Water Resources Systems and High-Quality Economic Growth in the Yangtze River Delta. *Water* **2023**, *15*, 4030.
10. Li, C.; He, L.; Liu, D.; Feng, Z. A Scientometric Review for Uncertainties in Integrated Simulation–Optimization Modeling System. *Water* **2024**, *16*, 285.

## References


1. Nagata, K.; Shoji, I.; Arima, T.; Otsuka, T.; Kato, K.; Matsubayashi, M.; Omura, M. Practicality of integrated water resources management (IWRM) in different contexts. *Int. J. Water Resour. Dev.* **2022**, *38*, 897–919. [CrossRef]
2. Chen, D.; Cai, Y.; Wang, X.; Li, C.; Yin, X.; Liu, Q. An inexact modeling approach for supporting water resources allocation under natural and social complexities in a border city of China and Myanmar. *Resour. Conserv. Recycl.* **2021**, *168*, 105245. [CrossRef]
3. Cai, X.; Wallington, K.; Shafiee-Jood, M.; Marston, L. Understanding and managing the food-energy-water nexus–opportunities for water resources research. *Adv. Water Resour.* **2018**, *111*, 259–273. [CrossRef]
4. Wang, X.; Chen, Y.; Li, Z.; Fang, G.; Wang, F.; Hao, H. Water resources management and dynamic changes in water politics in the transboundary river basins of Central Asia. *Hydrol. Earth Syst. Sci.* **2021**, *25*, 3281–3299. [CrossRef]
5. Jain, S.K.; Shilpa, L.S.; Rani, D.; Sudheer, K.P. State-of-the-art review: Operation of multi-purpose reservoirs during flood season. *J. Hydrol.* **2023**, *618*, 129165. [CrossRef]
6. Yu, X.; Xu, Y.-P.; Gu, H.; Guo, Y. Multi-objective robust optimization of reservoir operation for real-time flood control under forecasting uncertainty. *J. Hydrol.* **2023**, *620*, 129421. [CrossRef]
7. Zeng, Y.; Liu, D.; Guo, S.; Xiong, L.; Liu, P.; Chen, J.; Yin, J.; Wu, Z.; Zhou, W. Assessing the effects of water resources allocation on the uncertainty propagation in the water–energy–food–society (WEFS) nexus. *Agric. Water Manag.* **2023**, *282*, 108279. [CrossRef]

water

MDPI

*Article*

# Assessment of Hydrological and Meteorological Composite Drought Characteristics Based on Baseflow and Precipitation

**Saihua Huang [1,2], Heshun Zhang [1,2,*], Yao Liu [3], Wenlong Liu [4], Fusen Wei [4], Chenggang Yang [5], Feiyue Ding [6], Jiandong Ye [7], Hui Nie [1,2], Yanlei Du [8] and Yuting Chen [9]**

1 College of Hydraulic and Environmental Engineering, Zhejiang University of Water Resources and Electric Power, Hangzhou 310018, China; huangsh@zjweu.edu.cn (S.H.); nieh@zjweu.edu.cn (H.N.)
2 Key Laboratory for Technology in Rural Water Management of Zhejiang Province, Hangzhou 310018, China
3 Zhejiang Environmental Technology Co., Ltd., Hangzhou 310013, China; gracely95gh@163.com
4 Huzhou Hydrological Center, Huzhou 313000, China; goddragon116@163.com (W.L.); w17705820391@163.com (F.W.)
5 Ningbo Water Resources Information Management Center, Ningbo 315000, China; 13777129922@163.com
6 Shaoxing Shunjiangyuan Provincial Nature Reserve Management Center, Shaoxing 312000, China; chachuqiu0668@163.com
7 Tonglu County Jiangnan Irrigation District Project Management Office, Hangzhou 311500, China; tlyjd158@163.com
8 Changxing County Environmental Protection Monitoring Station, Huzhou 313100, China; duanlei007@126.com
9 Zhejiang Water Conservancy Development Planning Research Center, Hangzhou 310012, China; chenyt@zjwater.gov.cn
* Correspondence: 2021b07061@stu.zjweu.edu.cn

**Abstract:** Traditional univariate drought indices may not be sufficient to reflect comprehensive information on drought. Therefore, this paper proposes a new composite drought index that can comprehensively characterize meteorological and hydrological drought. In this study, the new drought index was established by combining the standardized precipitation index (SPI) and the standardized baseflow index (SBI) for the Jiaojiang River Basin (JRB) using the copula function. The prediction model was established by training random forests on past data, and the driving force behind the combined drought index was explored through the LIME algorithm. The results show that the established composite drought index combines the advantages of SPI and SBI in drought forecasting. The monthly and annual droughts in the JRB showed an increasing trend from 1991 to 2020, but the temporal characteristics of the changes in each subregion were different. The accuracies of the trained random forest model for heavy drought in Baizhiao (BZA) and Shaduan (SD) stations were 83% and 88%, respectively. Furthermore, the Local Interpretable Model-Agnostic Explanations (LIME) interpretation identified the essential precipitation, baseflow, and evapotranspiration features that affect drought. This study provides reliable and valid multivariate indicators for drought monitoring and can be applied to drought prediction in other regions.

**Keywords:** composite drought index; baseflow; LIME algorithm; Jiaojiang River Basin

## 1. Introduction

As a global problem, drought poses a significant challenge to the development of human societies. With the increase in global warming, the frequency of extreme weather events worldwide has increased, which has led to a gradual increase in the frequency and scope of droughts, posing a severe threat to agricultural production, economic development, and social stability. Although China has a vast area and many rivers, its uneven spatial distribution makes it a drought-prone country. According to statistics, the economic losses caused by meteorological disasters account for about 50% of all natural disasters, and the losses caused by droughts account for more than 50% of meteorological disasters [1]. To better understand and measure the characteristics of drought, scholars have adopted a variety of drought indices, such as meteorological and hydrological indices, to assess and compare droughts objectively. These drought indices not only help to measure the intensity, duration, and frequency of drought occurrence but also provide a more comprehensive assessment of its impacts on different areas and the degree of damage. Through the comprehensive analysis of different indicators, scholars can more accurately grasp the characteristics of drought and its impact on human society, which provides an important basis for formulating strategies and measures to deal with drought.

Researchers have constructed different drought evaluation indices for different types of droughts (meteorological drought, hydrological drought, agricultural drought, socio-economic drought). For example, the Standardized Precipitation Evapotranspiration Index (SPEI) [2] describes meteorological droughts, and the Standardized Streamflow Index (SSI) [3] investigates hydrological droughts. However, because of the differences in drought-inducing variables, a single drought index cannot comprehensively evaluate agricultural, meteorological, hydrological, and socio-economic aspects [4–6]. Creating a comprehensive drought index that includes multiple drought elements is necessary. Waseem et al. [7] proposed a new multivariate drought evaluation method using entropy-weighted Euclidean distance to create a composite drought index. Also, Huang et al. [8] combined SPI and SSI and constructed another composite drought index using the entropy weighting method in 2015. In addition, Liu et al. [9] proposed a comprehensive composite drought index by applying the principal component analysis (PCA) method, which combined precipitation, evapotranspiration, soil moisture, streamflow, and other hydrometeorological factors.

However, comprehensive drought indicators based on the assignment and fuzzy synthesis methods have a certain subjective tendency when assigning weights, which cannot process the data objectively and cause errors. The PCA method cannot reflect the nonlinear relationship among the relevant variables. The copula function is a kind of joint distribution that can construct the marginal distribution as an arbitrary distribution, which can effectively describe the correlation among variables and has a wide range of applications in hydrology and water resources [10–15]. Azhdari et al. [16] constructed three composite hydrometeorological indices, including JDHMI-CCA, JDHMI-PCA, and JDHMI-copula, using typical correlation analysis (CCA), principal component analysis (PCA), and copula-based methods, and explored the mechanism of linear and nonlinear methods in drought status assessment. Wang et al. [17] established a new meteorological and hydrological drought index (MSDIP) using streamflow and precipitation as indicators. However, factors such as precipitation intensity and duration can affect rapid changes in streamflow, making it relatively unstable. Baseflow is more stable than streamflow derived by long-term groundwater recharge, which is usually less affected by short-term meteorological changes [18]. Odongo et al. [19] selected possible wet and dry conditions in a study area based on SPI and SSI according to a weighted similarity metric. The essential baseflow characteristics to replenish river flow during dry periods, SBI instead of SSI, can better respond to hydrologic drought characteristics. Bazrkar and Chu [20] improved hydrological drought identification by developing a new standardized base flow index. Meresa et al. [21] explored the relationship between hydrological and meteorological drought and watershed characteristics. They used the SPI, SPEI, SRI, and SBI to study drought propagation under meteorological to hydrological drought conditions in nine selected catchment areas of the Awash Basin in Ethiopia. Kwak et al. [22] used the SBI to consider the dry and wet conditions of a dam and its storage capacity. In addition, many studies [23–25] have demonstrated that baseflow can be used as a hydrological characteristic factor to characterize hydrological drought conditions in a watershed. Therefore, this study constructs a comprehensive index based on the copula function by constructing a joint distribution function with precipitation and baseflow as marginal distributions. The SPI is used to characterize meteorological droughts and the SBI is used to characterize hydrological droughts so that the index can comprehensively characterize the joint features of meteorological droughts and hydrological droughts and improve the accuracy and applicability of the composite drought index.

Meanwhile, with the development of computer intelligence technology, machine learning models have been applied to many fields, and the high accuracy of such models has been proven. In meteorological and hydrological drought forecasting, the use of machine learning models to construct regional meteorological and hydrological drought forecasting models has made some progress. Compared with the traditional statistical regression methods, Support Vector Machine (SVM) [26], Back Propagation Neural Network (BPNN) [27], eXtreme Gradient Boosting (XGBoost) [28], random forest (RF) [29,30], and other machine learning algorithms have significant advantages in processing large-scale and multi-source remote sensing data and have been gradually used in meteorological and hydrological forecasting. For example, Elbeltagi et al. [31] investigated the prediction accuracy of SPI-based RF, Random Tree (RT), and Gaussian Process Regression (GPR-PUK kernel) models for forecasting meteorological droughts in semi-arid regions. Lotfirad et al. [32] used an RF model to demonstrate that the prediction accuracy of the SPI and SPEI can be improved by increasing the time scale. It was also demonstrated that the SPEI is more capable than the SPI in drought prediction. Zheng et al. [33] evaluated the predictive performance of random forest models combined with artificial intelligence on a test set of four stocks using optimal parameters. RF models are heavily used in several areas, such as hydrologic forecasting, finance, and the environment. It is a more mature forecasting model. In this paper, using the RF model to forecast regional drought ensures the accuracy of the prediction results and the significance of each drought-affecting variable, as explained by the subsequent LIME interpretation.

Therefore, the main objectives of this study are (1) to calculate the SPI and the SBI based on the optimal marginal distribution function using precipitation and baseflow data; (2) to introduce a new copula-based comprehensive drought index, which combines the advantages of the SPI and the SBI and can simultaneously reflect both meteorological and hydrological drought; (3) to assess the drought in the Jiaojiang River Basin (JRB) from 1991 to 2020 based on drought indices; and (4) to predict future droughts using a random forest model and to develop a complex machine learning algorithm for the prediction of droughts using the Local Interpretable Model-Agnostic Explanations (LIME) algorithm to interpret the results of making complex machine learning model predictions.

## 2. Materials and Methods

### 2.1. Study Area and Dataset

#### 2.1.1. Study Area

The Jiaojiang River Basin (JRB) is one of the eight important river basins in Zhejiang Province. The JRB is located on the central coast of Zhejiang Province between 120°17′6′′ and 121°41′00′′ east longitude and 28°32′2′′ and 29°20′29′′ north latitude, with a watershed area of 6603 km$^2$ (Figure 1). The topography of the JRB slopes from west to east. The west–central and northern Zhongshan Mountains are rolling, the coastal plain is embedded with low hills, and the river channels are densely scattered.

#### 2.1.2. Datasets

Twenty-two meteorological datasets and two hydrological datasets of observation records in the JRB were obtained from the National Climate Center of the National Meteorological Administration of China (Figure 1). The records cover hydrologic information from 1991 to 2020. In this study, the JRB was divided into two sub-basins for analysis, including the Yong'an Creek Basin (Shaduan station) and the Shifeng Creek Basin (Baizhiao station).

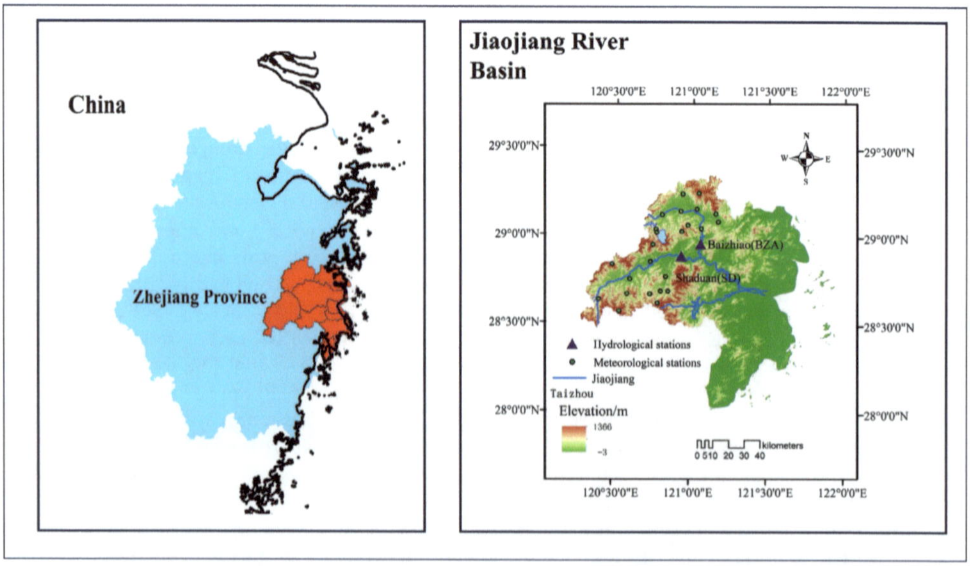


**Figure 1.** Spatial distribution of meteorological and hydrological stations in the JRB.

### 2.2. Methods

The SPI (meteorological drought) and the SBI (hydrological drought) were selected to construct a composite drought index using the Frank copula function to characterize meteorological and hydrological drought. A random forest model was trained on the data from 1991 to 2010 for the Shaduan (SD) and Baizhiao (BZA) stations to predict the composite drought index from 2010 to 2021. Finally, the LIME algorithm was used to analyze the composite drought index prediction results and interpret the importance of each input factor. The characteristics of meteorological droughts transformed into hydrological droughts were investigated. A flow chart of the methods used in this study is shown in Figure 2.

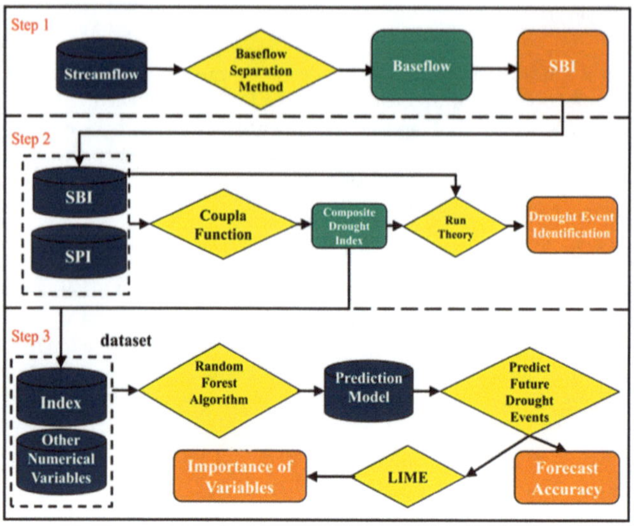


**Figure 2.** Flow chart of this study.

### 2.2.1. Drought Index

The SPI was chosen as the meteorological drought index, which can quantitatively characterize regional drought at different time scales [34]. The SPI considers that precipitation obeys a skewed distribution and normalizes precipitation, which is a good prospect for application [35,36]. The specific computational steps of the SPI are as follows:

$$f(x) = \frac{1}{\beta^{\gamma}\Gamma(\gamma)}x^{\gamma-1}e^{-x/\beta} \quad (1)$$

where $\beta > 0$ and $\gamma > 0$ are scale and shape parameters, respectively. $\beta$ and $\gamma$ are estimated by the linear method of moments. $\Gamma(\gamma)$ is used to fit the cumulative precipitation series for a given time scale.

The cumulative probability of the precipitation sequence x is:

$$\mathrm{F(x)} = \int f(x)dx \quad (2)$$

The standard normalization was performed as follows:

$$F(x < x_0) = \frac{1}{\sqrt{2\pi}}\int e^{-Z^2/2}dx \quad (3)$$

The final SPI value was obtained as follows:

$$SPI = Z = S\{t - \frac{(0.10328t + 0.802853)t + 2.515517}{[(0.001308t + 0.189269)t + 1.432788]t + 1}\} \quad (4)$$

where $t = \sqrt{\ln\frac{1}{F^2}}$, and when $F > 0.5$, $F = 0.5$, $S = 1$. When $F \leq 0.5$, $S = 1$.

The standardized baseflow index (SBI) was selected as the hydrological drought index. The baseflow was selected from the digital filtering method of the Chapman Maxwell (CM) [37] method for baseflow separation, which effectively solves the uncertainty problem of baseflow when surface streamflow stops flowing. The calculation formula is:

$$Q_{b(i)} = \frac{a}{2-a}Q_{b(i-1)} + \frac{1-a}{2-a}Q_i \quad (5)$$

where $Q_{b(i)}$ is the base flow rate at the moment, mm; $Q_{b(i-1)}$ is the base flow rate at the moment, mm; $Q_i$ is the streamflow flow rate at the moment, mm; $i$ is the time step, d; and $a$ is the receding water constant. The value can be obtained through a receding water analysis. Usually, the empirical value is set to 0.925, as obtained by Nathan and McMahon [38] based on the characterization of six watersheds in Germany and used for baseflow partitioning. Considering the uncertainty in the empirical recession constants when applied to other watersheds, estimating the recession constant values using the CM filtering method can reduce the uncertainty in baseflow separation and thus improve accuracy. In this study, we used the automated baseflow identification method developed by Cheng et al. [39] to compute the receding water constants of watersheds and perform baseflow separation.

The SBI calculation method is similar to the SRI computation since the baseflow value is acquired after the filtering method processes the streamflow data. The baseflow is utilized in place of the streamflow to perform normalization calculations, and the baseflow probability distribution type is chosen and standardized. The drought index is then split into two categories, as indicated in Table 1.

**Table 1.** Labeling of the aridity index divisions.

| Degree of Aridity | SPI Value | SBI Value | Composite Drought Index |
|---|---|---|---|
| No drought | $SPI \geq -0.5$ | $SBI \geq -0.5$ | $CDI \geq -0.5$ |
| Normal drought | $-1 \leq SPI < -0.5$ | $-1 \leq SBI < -0.5$ | $-1 \leq CDI < -0.5$ |
| Heavy drought | $SPI < -1$ | $SBI < -1$ | $CDI < -1$ |

2.2.2. Run Theory

In this study, we choose the run theory [40] for the initial judgment of the drought index time series, rejecting and fusing three-step identification to extract drought features. The three thresholds set are $R_0 = 0, R_1 = -0.5$, and $R_2 = -1$. The specific steps are as follows: (1) Initial identification when the drought index is less than $R_1$, then the initial judgment of the month for drought is made, as shown in Figure 3a. (2) Elimination of non-droughts, when the drought index of a month is between $R_1$ and $R_2$, and only this month reaches this standard, it is recognized that the month is non-drought and is to be eliminated, as shown in Figure 3*b*.

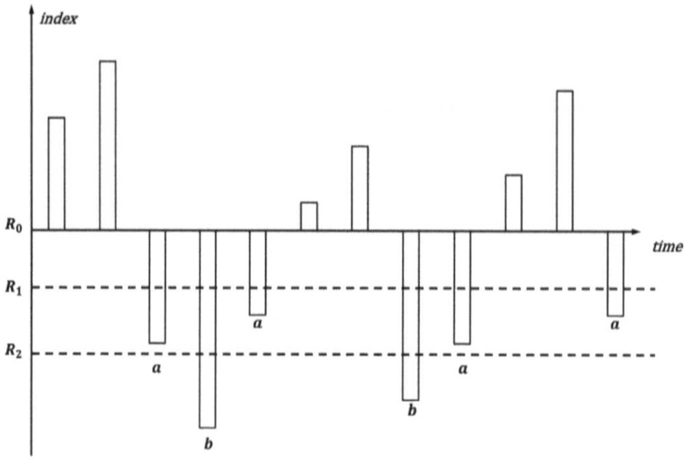


**Figure 3.** Schematic diagram of run theory. $R_0$, $R_1$, and $R_2$ are the drought index equal to 0, −0.5, and −1, respectively.

2.2.3. Theory of Two-Dimensional Copula Functions

The copula function is an important statistical tool for describing the distributional relationships of multidimensional random variables. It is mainly used to model the joint distribution function of random variables, which separates the marginal distribution from dependence and makes the model more flexible. There are various types of copula functions, and the common ones include Gaussian copula [41], t-distribution copula [42], Clayton copula [43], and Frank copula [44]. Each type of copula corresponds to a different dependence structure, so choosing an appropriate copula function is crucial for accurately describing the dependence among variables. Among them, the Frank copula function has a simple structure. It can be used to describe the symmetric correlation structure, which applies to both positive and negative correlations and has no limitation on the degree of correlation, and the variation in the upper and lower tails of the correlations is not apparent [45]. Many drought studies have used the Frank copula function [46–48]. Therefore, this study chooses the Frank copula connecting function to construct the joint distribution of precipitation and baseflow.

According to Sklar's theorem, let F and G be the marginal distribution functions of the random variables x and y, respectively, and H be the joint distribution function, then $\forall x, y \in \overline{R}$ have copula function C such that:

$$H(x,y) = C(F(x), G(y)) \tag{6}$$

If $F$ and $G$ are continuous, then C is unique.

The function was first proposed by Frank in 1979, and its expression is:

$$C(\mu, v) = -\frac{1}{\theta}\ln[1 + \frac{(e^{-\theta\mu} - 1)(e^{-\theta v} - 1)}{e^{-\theta} - 1}], \theta \in R \tag{7}$$

The generated meta is as follows:

$$\phi_\theta(t) = \ln[\frac{(e^{-\theta t} - 1)}{e^{-\theta} - 1}], \theta \in R \tag{8}$$

where $\mu$ and $v$ are the marginal cumulative probabilities of the two variables, respectively, and $\theta$ is a parameter. It can be obtained from the Kendal rank correlation coefficient $\tau$:

$$\tau = 1 - \frac{4}{\theta}(D_1(-\theta) - 1) \tag{9}$$

The first-order Debye function $D_1(\theta)$ expression is:

$$D_1(\theta) = \frac{1}{\theta}\int_0^\theta \frac{t}{e^t - 1}dt \tag{10}$$

2.2.4. Establishment of A Composite Drought Index

The precipitation and baseflow of the BZA and SD stations in the JRB are random variables $X$ and $Y$, respectively. Through the Pearson correlation coefficient method, the correlation coefficients of precipitation and streamflow in each region were calculated to be 0.7884 and 0.7506, which have solid correlations and can be used to construct the joint distribution function. Where $X$ and $Y$ represent a particular value of precipitation and baseflow in the plain and sandy section of Pak Chi, assuming that the corresponding marginal distributions of the two random variables are $F(x)$ and $G(y)$, their joint distribution $P$ can be expressed by the cumulative joint probability $p$ and copula function C as:

$$P(x \le X, y \le Y) = C[F(x), G(y)] = p \tag{11}$$

Thus, the composite drought index is obtained from the joint distribution function:

$$Index = \phi^{-1}(p) \tag{12}$$

where $\phi$ is the standard normal distribution.

Since meteorological droughts begin and end relatively quickly (because of a lack of precipitation), hydrological droughts (because of insufficient streamflow) begin and end with some delay in response to meteorological droughts. It is often the case that a meteorological drought has already ended, and a hydrological drought has just begun. This situation makes it difficult for decision-makers to formulate timely, rational, and effective response strategies. The composite drought index (CDI) is a composite index that characterizes both meteorological and hydrological droughts. It combines baseflow and precipitation data to capture drought onset, the severity of drought conditions, and drought duration. In this study, a similar approach to the SPI classification was used to categorize the drought classes of the CDI, as shown in Table 1.

The CDI is particularly effective because it considers meteorological and hydrological factors. Meteorological droughts are characterized by below-average precipitation, while hydrological droughts are characterized by below-average surface water and groundwater levels. The CDI captures both factors and comprehensively describes drought conditions. The CDI is an essential tool for drought monitoring and management. Identifying different levels of drought severity can help policymakers and water managers make informed decisions on water allocation and conservation measures. In addition, the CDI can be used to track changes in drought conditions over time, thus providing early warning of potential drought emergencies. Overall, the CDI represents a significant advance in drought monitoring and management. It captures meteorological and hydrological factors to provide a comprehensive picture of drought conditions and is a composite of meteorological and hydrological indices [49].

#### 2.2.5. Linear Regression Estimator

The linear regression method is used to estimate the slope. A positive slope value indicates an increasing trend, while a negative value indicates a decreasing trend. The linear regression line can be computed as follows:

$$y = ax + b \tag{13}$$

where $x$ and $y$ are the explanatory variable and the dependent variable, respectively, while $a$ and $b$ are the slope and intercept, respectively.

#### 2.2.6. Mann–Kendall Trend Test

Most previous studies have assumed that sample data are serially independent. However, it is known that some hydrometeorological time series, such as water quality and flow or rainfall time series, may exhibit serial correlation. In such cases, serial correlation affects the ability of the Mann–Kendall (MK) [50] test to assess the significance of a trend because Mann–Kendall and Theil–Sen cannot consider the AR (1) process for the time series.

The nonparametric MK test is the most widely used time series trend detection method. If the total amount of data in the time series is denoted by $N$, the statistic $S$ can be calculated:

$$S = \sum_{i=1}^{N-1} \sum_{j=i+1}^{N} sgn(Y_j - Y_i) \tag{14}$$

If the total amount of data in the time series is denoted by $N$, then the statistic is as follows:

$$sgn(\theta) = \begin{cases} +1 & if\ \theta = Y_j - Y_i > 0 \\ 0 & if\ \theta = Y_j - Y_i = 0 \\ -1 & if\ \theta = Y_j - Y_i < 0 \end{cases} \tag{15}$$

A positive (negative) value of $S$ indicates an upward (downward) trend. $S$ is considered to be normally distributed when $N \geq 8$, and its mean and variance can be computed as follows:

$$E[S] = 0 \tag{16}$$

$$var = \frac{[N(N-1)(2N+5) - \sum_{i=1}^{n} t_i i(i-1)(2i+5)]}{18} \tag{17}$$

where $t_i$ indicates the amount of data in the $i$th tied group. Finally, the standardized test statistics $Z$ can be computed as follows:

$$Z = \begin{cases} (S-1)/\sqrt{var(S)} & S > 0 \\ 0 & S = 0 \\ (S+1)/\sqrt{var(S)} & S < 0 \end{cases} \tag{18}$$

A positive value of $Z$ indicates an increasing trend, while a negative value indicates a decreasing trend. In this study, trends were tested with a significance level of $\alpha$ = 0.05.

The null hypothesis of no trend was rejected if the absolute value of $Z$ was more significant than 1.96.

2.2.7. Random Forest

A random forest (RF) method was used to predict future drought conditions in the JRB, based on an algorithm that Breiman proposed [51]. The method constructs a decision tree for each sample by randomly selecting subsamples from the original sample multiple times in a relaxed manner [52,53]. Then, the prediction results of multiple decision trees are combined by averaging or voting to determine the final prediction result. Elbeltagi et al. [31] predicted the accuracy of meteorological drought in semi-arid regions using the standardized precipitation index (SPI) based on random forest (RF), random tree (RT), and Gaussian process regression (GPR-PUK kernel) models. Zarei et al. [54] compared the ability of six more commonly used drought indices in agricultural drought assessment using the RF algorithm.

The composite drought index in this study was predicted using a typology to make predictions for different levels of drought. In categorical analytics research, the complex black-box nature makes it difficult for researchers to make sense of data insights and model decisions [55]. Therefore, this study employs the LIME algorithm to interpret the predicted results using LIME (Figure 4). Data storytelling, as a practical process with an explanatory purpose, can help storytelling audiences understand the context and reasons for generating model predictions, enabling an immersive user experience and deep cognitive data. Introducing interpretability techniques into the data storytelling process can enhance the explanatory effect of data stories, which can assist in decision-making.

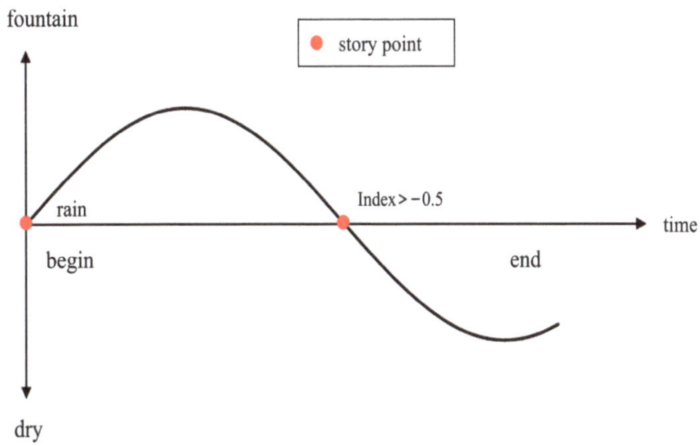


**Figure 4.** Structure of the drought story narrative.

Interpretable machine learning techniques offer a fresh approach to locating the data story reversal point in the context of classification model prediction outcomes. Using interpretable machine learning techniques, we may identify the critical components in the decision-making process and better understand how a model makes a particular prediction. This enables us to pinpoint pivotal moments within the data narrative, representing significant characteristics or patterns that lead the algorithm to alter its forecasts.

## 3. Results

### *3.1. Comprehensive Hydrometeorological Drought Characteristics*

3.1.1. Establishment of Composite Index and Analysis of Its Trends

According to the streamflow of the BZA and SD hydrological stations in the JRB from 1991 to 2020, the baseflow of the two stations was divided according to Equation (5). Figure 5 shows that the fluctuation intensities in the baseflow after the division of the two places are significantly reduced compared with the streamflow, indicating that the baseflow is more stable than the streamflow, which is in line with the needs of this paper. Subsequently, the SBI indices of the two sections were calculated based on the divided torrent data. First, the hydrological drought index (SBI) and meteorological drought index (SPI) were computed in this study. Then, a novel drought index based on copula was suggested, capable of capturing both hydrological and meteorological drought. After that, the superiority of the composite drought index was confirmed by contrasting the SPI, SBI, and CDI on a monthly scale. The Frank copula parameter values for the BZA and SD basins were determined to be 5.5160 and 4.1631, respectively.

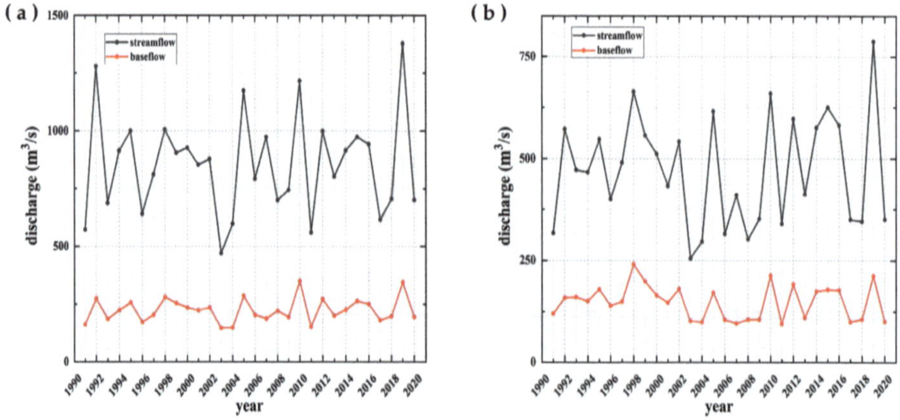


**Figure 5.** Annual baseflow and streamflow from the (**a**) BZA and (**b**) SD stations.

Figure 6 shows the SPI, SBI, and CDI of the JRB at the annual scale from 1991 to 2021. The results of the linear fit for the CDI, SPI, and SBI on the annual scale are shown in Figure 6. The results of the MK test are shown in Table 2. It can be found that the results of the MK test are the same as those of the linear fitting, which proves the accuracy of the trend change. The Pearson correlation coefficients of the composite drought index with the SPI and SBI values passed the 95% confidence test, which indicates that the newly constructed composite drought index has some reliability [56]. Figure 6 shows that the aridity in the BZA area decreases year by year, but the aridity in the SD area increases year by year. According to Figure 6, the fluctuation shapes of the CDI, SPI, and SBI are similar, and when the SPI and SBI decrease, the CDI decreases at the same time, and when the SPI and SBI increase, the CDI increases at the same time, which indicates that the new CDI has very high accuracy and reliability in recognizing the two kinds of droughts, namely, meteorological drought (SPI) and hydrological drought (SBI). The trend line in Figure 6 indicates that when the SPI and SBI rise simultaneously, the CDI likewise rises simultaneously. Similarly, the CDI falls when the SPI rises and the SBI falls, but its slope is much smaller than the SBI. This indicates that the new CDI combines the two types of drought indices rather than considering them as one single drought index, which can be used to recognize drought more comprehensively.

**Table 2.** Results of the MK test.

| Station (Drought Index) | Slope | Z |
|---|---|---|
| SD(SPI) | 0.0011 | 2.0561 |
| SD(SBI) | −0.0015 | −2.7836 |
| SD(CDI) | −0.0002 | −0.4896 |
| BZA(SPI) | 0.0011 | 2.028 |
| BZA(SBI) | 0.0007 | 1.417 |
| BZA(CDI) | 0.0011 | 1.9859 |

To make a more visual comparison, we intercepted the changes in the monthly SPI, SBI, and CDI (1992–1998), as shown in Figure 7. Some differences in the three indices can be observed when zooming in on the abscissa. From the inter-annual variation in the index in Figure 7, it can be seen that droughts occurred more frequently and at higher drought levels from 1992 to about 1998, and the same was valid from 1999 to about 2020. The droughts captured by the drought index in this study are temporally consistent with the droughts that historically occurred in the JRB [57], which suggests that the newly constructed drought index has some accuracy. The CDI based on the Frank copula function constructed in this paper has the advantages of reliability, sensitivity, and comprehensiveness. It can identify droughts more comprehensively and issue early warnings, supporting drought detection and prevention.

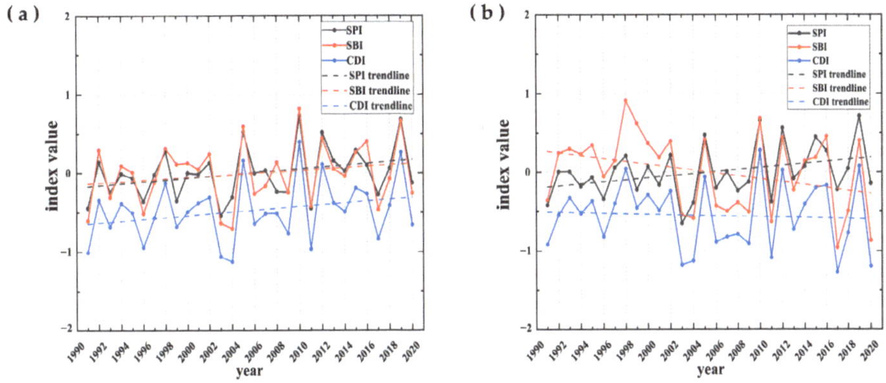


**Figure 6.** The annual SPI, SBI, and CDI at the (**a**) BZA and (**b**) SD stations.

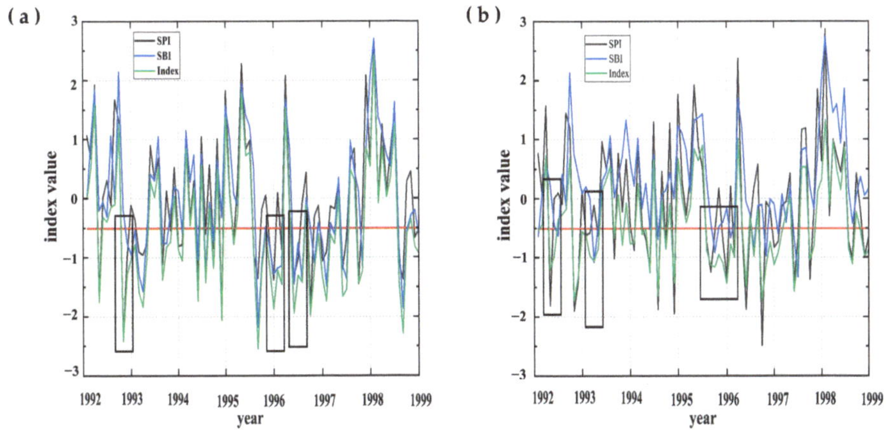


**Figure 7.** The monthly SPI, SBI, and CDI at the (**a**) BZA and (**b**) SD stations. The red line is the drought identification line.

#### 3.1.2. Mechanisms Underlying the Propagation of Meteorological Drought to Hydrological Drought

The three thresholds are $R_0 = 0, R_1 = -0.5$, and $R_2 = -1$ following the three-threshold approach of the tour theory for a drought index time series for the first judgment, culling, and fusion of the three-step identification and extraction of drought characteristics. Figure 8 displays the outcomes from 1995 to 1998. Concurrently, the tour theory was employed to better comprehend the modifications in meteorological and hydrological drought attributes following their dissemination. This aided in the identification of hydrological and meteorological drought occurrences, the extraction of drought intensity and ephemeral features, and the analysis of the outcomes. The results showed that the duration of meteorological and hydrological drought in the BZA station, as evaluated by the SPI and SBI, was 77 and 97 times, respectively. The meteorological droughts and hydrological droughts in the SD station were 74 and 83 times, respectively. According to the CDI results, the drought duration in the two stations was 150 and 150 times, respectively. Accordingly, the accuracy of the composite drought index in the two stations was 75.3% and 73.3%, respectively. Additionally, it may be inferred that hydrological droughts in the SD and BZA regions occurred one month and 0.7 months behind meteorological droughts.

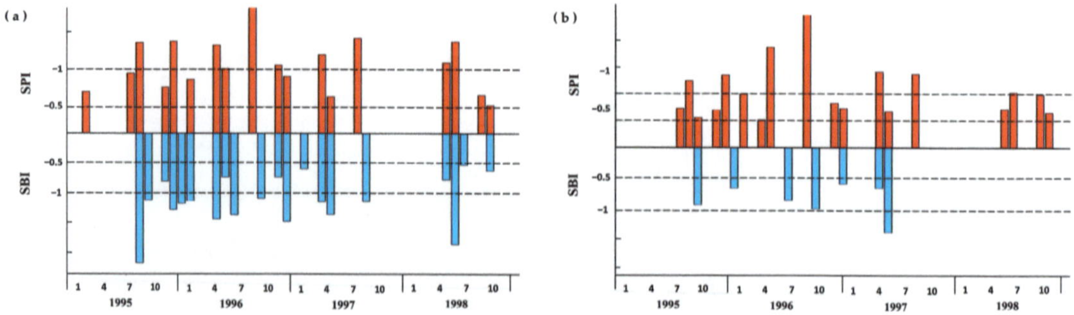


**Figure 8.** Meteorological and hydrological drought: (**a**) BZA basin and (**b**) SD basin.

## 3.2. *Future Drought Prediction and Important Factor Identification*

### 3.2.1. Drought Forecasting Performance

This study was based on the precipitation, baseflow, evaporation, El Niño Southern Oscillation (ENSO), Atlantic Multidecadal Oscillation (AMO), Arctic Oscillation (AO), North Atlantic Oscillation (NAO), Pacific Decadal Oscillation (PDO), Southern Oscillation (SO), Indian Ocean Dipole (IOD), and composite drought index (CDI) at the BZA and SD stations in the JRB during the period of 1991–2010. The composite drought index was also predicted for 2011–2020. The results of the correlation calculation among the numerical variables of the random forest model are shown in Figure 9. The correlation between the SO and ENSO indices was more significant than 0.9 at the BZA and SD stations. The correlation between the ENSO index and the other numerical variables was greater than that of the SO index. The accuracy of the exercise results is shown in Table 3. Based on the results, it can be seen that the trained model has vital prediction accuracy for no drought and heavy drought, but the prediction accuracy for normal drought still needs to be strengthened.

**Table 3.** Prediction accuracy of the random forest models.

| Categorization | BZA | SD |
| --- | --- | --- |
| No drought | 83% | 75% |
| Normal drought | 52% | 63% |
| Heavy drought | 83% | 88% |

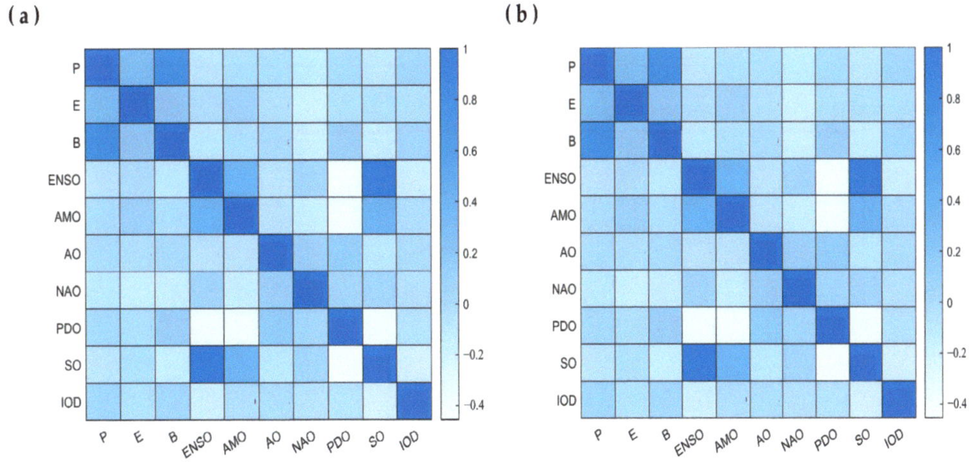


**Figure 9.** Correlation among indices in the random forest model at the (**a**) BZA and (**b**) SD stations.

### 3.2.2. Important Factor Identification

The inversion point identification algorithm and implementation scheme proposed in this paper apply to the subtype prediction results. For example, in drought forecasting, the data values are compared with the thresholds of −0.5 and −1. An area has no drought when the prediction result is more than −0.5. An area is in a normal drought when the result is more than −1 and less than −0.5. When the prediction value is less than −1, an area is in a heavy drought or above. This meets the needs of drought forecasting and supports the inversion point identification algorithm for validating data storytelling.

The algorithm model was trained using various information characteristics such as precipitation, baseflow, evapotranspiration, and the ENSO, AMO, AO, NAO, PDO, SO, and IOD indices. The LIME algorithm model was then utilized to predict the classification of drought, and the results were fed back to the decision maker for decision-making on drought conditions. The classification results were characterized as "index $\geq -0.5$" (no drought), "$-1 \leq$ index $< -0.5$" (normal drought), and "index $< -1$" (heavy drought and above).

Drought occurs in an area when the index is less than −0.5, which indicates a reversal of the need for change and indicates that the characteristics of the value may alter the outcomes. When the composite regional drought index is less than −0.5, the features of the value can be altered. Each time a variable's properties change inside the tale's structure, this can be considered data storytelling in the story point narrative. Figure 10 shows this process. In the development stage of data storytelling, which runs from story point $F_0$ to $F_2$, the narrative progressively builds to a climax as one or more crucial feature variables are changed. Based on the LIME algorithm, data storytelling interpretation can be characterized in the following ways, as shown in Figure 10. The decision maker continuously looks for the reversal moment, or the point in the story where the predictions made by the model are overturned by altering the values of one or more features with higher weights. This is performed using the LIME method to output the feature weights for each prediction. Figure 11 displays the characteristic weights for every predicted outcome. Precipitation, baseflow, and evaporation are the fundamental eigenvalues influencing the impact of a drought.

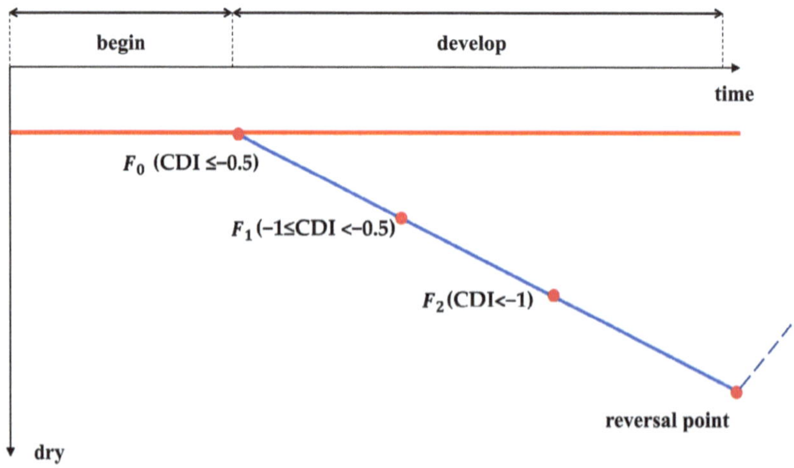


**Figure 10.** Inversion point identification and process as oriented by LIME algorithm interpretation. The red line is the drought judgment line, and the blue line is the drought process line.

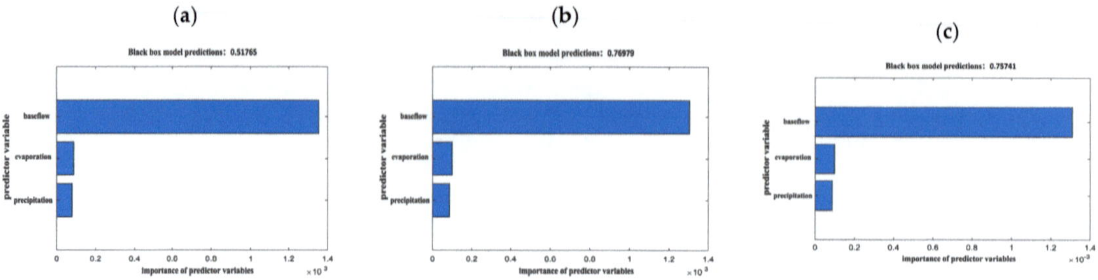


**Figure 11.** The importance of various variables for forecasting drought. The feature weights for (**a**) no drought, (**b**) normal drought, and (**c**) heavy drought prediction.

## 4. Discussion

Many new drought indices have been developed in past studies to monitor drought [58–60]. However, no comprehensive drought index is based on the copula function that combines the SPI and SBI to construct a comprehensive drought index that simultaneously characterizes meteorological and hydrological droughts. On this basis, a comprehensive drought index based on the Frank copula is proposed to characterize JRB droughts comprehensively. The trend line in Figure 6 reveals that the meteorological drought in the sand section has decreased yearly, but the hydrological drought has increased yearly. The reason for this phenomenon may be the faulty management of reservoirs because of the intensification of extreme climate change caused by global warming. As shown in Figure 7, the composite drought index can monitor the occurrence of meteorological drought, hydrological drought, or both. In general, the onset of a drought is usually due to a persistent lack of rainfall, so the SPI index is susceptible to capturing the onset of a drought. There is a delay in the response of hydrological droughts to meteorological droughts because of complex sink-producing processes, so the SBI effectively identifies the duration and end of droughts.

As shown in the first black rectangular boxes in Figure 7a,b, the moments when the CDI and SPI values are less than $-0.5$ (considered to be the beginning of the drought) are earlier relative to the SPI, which suggests that the CDI was comparable to the SPI in its ability to capture the beginning of the drought. In addition, the composite drought index appears to be similar to the SBI in terms of its capacity to detect the end of drought because the times at which the CDI and the SBI are more significant than $-0.5$ (the point at which the drought is deemed to be over) occur later than the SPI. This is because the SPI was mainly based on meteorological drought characteristics, which develop suddenly and rapidly. The SBI was based on hydrological drought characteristics, with considerable delays and longer durations. The third black rectangles in Figure 7a,b show that the CDI can include both the beginning and the end of a drought, as captured by the SPI and the SBI. The CDI captures the beginning of the SPI and the duration of the SBI during drought events. Furthermore, as demonstrated by the variations in the three indices in the second black rectangular boxes in Figure 7a,b, the CDI accurately depicted the beginning of the drought in both cases when only a hydrological drought occurred and when no meteorological drought occurred. These findings offer compelling evidence that the composite drought index, which incorporates data on streamflow and precipitation, can accurately and sensitively identify the beginning, middle, and end of a drought.

Therefore, we believe that the proposed composite index has practical applications. Once the CDI detects a drought, practitioners, watershed managers, or agencies should remain vigilant in warning and preventing possible droughts. In addition, the CDI was significantly correlated with the SPI and the SBI ($p < 0.05$), indicating that the CDI has good reliability. In addition, Hao and AghaKouchak [61] also compared the new composite drought index with the SPI and the standardized soil moisture index by analyzing the drought onset and propagation time they captured. They verified that the copula-based composite drought index is reliable and has effective drought early warning capability. In Figure 7, the green line representing the composite drought index is slightly lower than the other lines, indicating that the conditions captured by the composite drought index are more severe than those captured by the single-factor drought index. Without accurate forecasts, people prefer predicted drought levels to be higher than actual levels, which is more conducive to drought planning.

The number of hydrologic droughts at BZA is significantly higher than at SD, as shown in Figure 8. This may be because the SD station controls 75% of the watershed, whereas the BZA station controls only 40%. In addition, a random forest model was trained on precipitation, baseflow, evapotranspiration, and the ENSO, AMO, AO, NAO, PDO, SO, and IOD indices, which had a more accurate accuracy for no drought and heavy drought conditions. The LIME model was then used to interpret the predictions of the random forest to identify the leading causes of drought in the JRB region. The results showed that precipitation, baseflow, and evapotranspiration most significantly impact the JRB regional drought. This is consistent with the fact that precipitation, baseflow, and evapotranspiration are all determinants of drought in each region and that the JRB is a large and stable watershed, so it is less affected by factors such as ESNO, AMO, AO, NAO, PDO, SO, and IOD.

There are some shortcomings in this study including the following: (1) only meteorological and hydrological droughts are considered, and agricultural and socio-economic droughts are excluded; (2) we validated the applicability of the Frank copula function in the JRB but failed to establish a generalized method for constructing a composite drought index; (3) there is still a need to improve the prediction accuracy of the random forest model trained in this paper for predicting mild droughts compared to other scenarios; and (4) we did not categorize the situation after a moderate drought, resulting in a lack of research on the composite drought index for extreme drought situations. Furthermore, the propagation mechanism from meteorological drought to hydrological drought was only considered on the time scale. Therefore, the spatial propagation mechanism will be further studied.

## 5. Conclusions

This study constructs a new comprehensive drought index based on the copula function according to precipitation and baseflow, which can better characterize meteorological and hydrological droughts. The random forest algorithm was used to predict drought, and the LIME algorithm was used to explain the importance of predictors. The following conclusions were obtained:

(1) The CDI can comprehensively characterize meteorological and hydrological droughts, providing reliable and powerful support for monitoring, prevention, control, and forecasting.

(2) From 1991 to 2020, the monthly and annual droughts in the BZA area of the JRB region showed a decreasing trend, and the monthly and annual droughts in the SD area showed an increasing trend. The subregions have different time-varying characteristics. Most droughts occurred in the fall and winter, with mean CDI values of −0.447 and −0.548 at the BZA and SD stations, respectively.

(3) The random forest model training results were accurate for no drought and heavy drought, with average validation accuracies greater than 0.8 at the BZA and SD stations. The LIME interpretation was used to find out the most important values of the characteristics of drought impacts, such as precipitation, baseflow, and evaporation.

**Author Contributions:** Conceptualization, H.Z.; methodology, W.L. and C.Y.; software, F.W. and F.D.; validation, Y.L.; formal analysis H.N.; investigation, S.H.; resources, H.N. and Y.D.; data curation, J.Y. and Y.C.; writing—original draft preparation H.Z.; writing—review and editing, S.H. All authors have read and agreed to the published version of the manuscript.

**Funding:** This research was funded by the Zhejiang Provincial Water Resources Department Science and Technology Program, grant number RC2201 and RB2002; Huzhou Science and Technology Plan Project, grant number 2023GZ64; the Natural Science Foundation of Zhejiang Province, grant number LZJWY22E090007; and the Scientific Research Fund of Zhejiang Provincial Education Department, grant number Y202352492.

**Data Availability Statement:** The data presented in this study are available on request from the corresponding author. The data are not publicly available due to [data are being used in another ongoing research].

**Acknowledgments:** Thanks to the Zhejiang Provincial Hydrological Management Center for providing the necessary hydrological and meteorological data for our research.

**Conflicts of Interest:** Author Yao Liu was employed by the company Zhejiang Environmental Technology Co., Ltd. The remaining authors declare that the research was conducted in the absence of any commercial or financial relationships that could be construed as a potential conflict of interest.

## References

1. Ayantobo, O.O.; Li, Y.; Song, S.; Javed, T.; Yao, N. Probabilistic modelling of drought events in China via 2-dimensional joint Copula. *J. Hydrol.* **2018**, *559*, 373–391. [CrossRef]
2. Oloruntade, A.J.; Mohammad, T.A.; Ghazali, A.H.; Wayayok, A. Analysis of meteorological and hydrological droughts in the Niger-South Basin, Nigeria. *Glob. Planet. Chang.* **2017**, *155*, 225–233. [CrossRef]
3. Marcos-Garcia, P.; Lopez-Nicolas, A.; Pulido-Velazquez, M. Combined use of relative drought indices to analyze climate change impact on meteorological and hydrological droughts in a Mediterranean basin. *J. Hydrol.* **2017**, *554*, 292–305. [CrossRef]
4. Salvadori, G.; De Michele, C. Multivariate real-time assessment of droughts via copula-based multi-site Hazard Trajectories and Fans. *J. Hydrol.* **2015**, *526*, 101–115. [CrossRef]
5. Azam, M.; Maeng, S.J.; Kim, H.S.; Lee, S.W.; Lee, J.E. Spatial and temporal trend analysis of precipitation and drought in South Korea. *Water* **2018**, *10*, 765. [CrossRef]
6. Huang, S.; Huang, Q.; Chang, J.; Zhu, Y.; Leng, G.; Xing, L. Drought structure based on a nonparametric multivariate standardized drought index across the Yellow River basin, China. *J. Hydrol.* **2015**, *530*, 127–136. [CrossRef]
7. Waseem, M.; Ajmal, M.; Kim, T.W. Development of a new composite drought index for multivariate drought assessment. *J. Hydrol.* **2015**, *527*, 30–37. [CrossRef]
8. Huang, S.; Chang, J.; Leng, G.; Huang, Q. Integrated index for drought assessment based on variable fuzzy set theory: A case study in the Yellow River basin, China. *J. Hydrol.* **2015**, *527*, 608–618. [CrossRef]
9. Liu, Y.; Zhu, Y.; Ren, L.; Yong, B.; Singh, V.P.; Yuan, F.; Jiang, S.; Yang, X. On the mechanisms of two composite methods for construction of multivariate drought indices. *Sci. Total Environ.* **2019**, *647*, 981–991. [CrossRef]
10. Kao, S.C.; Govindaraju, R.S. A copula-based joint deficit index for droughts. *J. Hydrol.* **2010**, *380*, 121–134. [CrossRef]
11. Lee, T.; Modarres, R.; Ouarda, T.B. Data-based analysis of bivariate copula tail dependence for drought duration and severity. *Hydrol. Process.* **2013**, *27*, 1454–1463. [CrossRef]
12. Maeng, S.J.; Azam, M.; San Kim, H.; Hwang, J.H. Analysis of changes in spatio-temporal patterns of drought across South Korea. *Water* **2017**, *9*, 679. [CrossRef]

13. Van de Vyver, H.; Van den Bergh, J. The Gaussian copula model for the joint deficit index for droughts. *J. Hydrol.* **2018**, *561*, 987–999. [CrossRef]
14. Meimandi, J.B.; Bazrafshan, O.; Esmaeilpour, Y.; Zamani, H.; Shekari, M. Risk assessment of meteo-groundwater drought using copula approach in the arid region. *Stochastic Environ. Res. Risk Assess.* **2024**, *38*, 1523–1540. [CrossRef]
15. Bazrafshan, O.; Zamani, H.; Mozaffari, E.; Azhdari, Z.; Shekari, M. Trivariate risk analysis of meteorological drought in Iran under climate change scenarios. *Meteorol. Atmos. Phys.* **2023**, *135*, 52. [CrossRef]
16. Azhdari, Z.; Bazrafshan, O.; Zamani, H.; Shekari, M.; Singh, V.P. Hydrometeorological drought risk assessment using linear and nonlinear multivariate methods. *Phys. Chem. Earth, Parts A/B/C.* **2021**, *123*, 103046. [CrossRef]
17. Wang, F.; Wang, Z.; Yang, H.; Di, D.; Zhao, Y.; Liang, Q. A new copula-based standardized precipitation evapotranspiration streamflow index for drought monitoring. *J. Hydrol.* **2020**, *585*, 124793. [CrossRef]
18. Chen, H.; Huang, S.; Xu, Y.-P.; Teegavarapu, R.S.V.; Guo, Y.; Nie, H.; Xie, H. Using baseflow ensembles for hydrologic hysteresis characterization in humid basins of Southeastern China. *Water Resour. Res.* **2024**, *60*, e2023WR036195. [CrossRef]
19. Odongo, R.A.; De Moel, H.; Van Loon, A.F. Propagation from meteorological to hydrological drought in the Horn of Africa using both standardized and threshold-based indices. *Nat. Hazards Earth Syst. Sci.* **2023**, *23*, 2365–2386. [CrossRef]
20. Bazrkar, M.H.; Chu, X. New standardized base flow index for identification of hydrologic drought in the Red river of the North Basin. *Nat. Hazards Rev.* **2020**, *21*, 05020011. [CrossRef]
21. Meresa, H.; Zhang, Y.; Tian, J.; Faiz, M. Understanding the role of catchment and climate characteristics in the propagation of meteorological to hydrological drought. *J. Hydrol.* **2023**, *617*, 128967. [CrossRef]
22. Kwak, J.; Joo, H.; Jung, J.; Lee, J.; Kim, S.; Kim, H.S. A case study: Bivariate drought identification on the Andong dam, South Korea. *Stoch. Environ. Res. Risk Assess.* **2021**, *35*, 549–560. [CrossRef]
23. Clancy, K.A. Standardized Baseflow Drought Index Comparison to SPEI in High Baseflow Streams. *J. Water Resour. Protect.* **2023**, *15*, 557–580. [CrossRef]
24. Porhemmat, J.; Altafi Dadgar, M. Analysis of hydrological drought indices in Alpine Zagros Mountains of Iran. *Arab. J. Geosci.* **2023**, *16*, 594. [CrossRef]
25. Lee, S.; Ajami, H. Comprehensive assessment of baseflow responses to long-term meteorological droughts across the United States. *J. Hydrol.* **2023**, *626*, 130256. [CrossRef]
26. Wang, H.; Li, G.; Wang, Z. Fast SVM classifier for large-scale classification problems. *Inform. Sci.* **2023**, *642*, 119136. [CrossRef]
27. Chen, L.; Wu, T.; Wang, Z.; Lin, X.; Cai, Y. A novel hybrid BPNN model based on adaptive evolutionary Artificial Bee Colony Algorithm for water quality index prediction. *Ecol. Indic.* **2023**, *146*, 109882. [CrossRef]
28. Piraei, R.; Afzali, S.H.; Niazkar, M. Assessment of XGBoost to estimate total sediment loads in rivers. *Water Resour. Manag.* **2023**, *37*, 5289–5306. [CrossRef]
29. Danandeh Mehr, A. A gene-random forest model for meteorological drought prediction. *Pure Appl. Geophys.* **2023**, *180*, 2927–2937. [CrossRef]
30. Hu, J.; Szymczak, S. A review on longitudinal data analysis with random forest. *Brief. Bioinf.* **2023**, *24*, bbad002. [CrossRef]
31. Elbeltagi, A.; Pande, C.B.; Kumar, M.; Tolche, A.D.; Singh, S.K.; Kumar, A.; Vishwakarma, D.K. Prediction of meteorological drought and standardized precipitation index based on the random forest (RF), random tree (RT), and Gaussian process regression (GPR) models. *Environ. Sci. Pollut. Res.* **2023**, *30*, 43183–43202. [CrossRef]
32. Lotfirad, M.; Esmaeili-Gisavandani, H.; Adib, A. Drought monitoring and prediction using SPI, SPEI, and random forest model in various climates of Iran. *J. Water Clim. Chang.* **2022**, *13*, 383–406. [CrossRef]
33. Zheng, J.; Xin, D.; Cheng, Q.; Tian, M.; Yang, L. The Random Forest Model for Analyzing and Forecasting the US Stock Market in the Context of Smart Finance. *arXiv* **2024**. [CrossRef]
34. McKee, T.B.; Doesken, N.J.; Kleist, J. The relationship of drought frequency and duration to time scales. In Proceedings of the 8th Conference on Applied Climatology, Anaheim, CA, USA, 17–22 January 1993; Volume 17, pp. 179–183.
35. Vicente-Serrano, S.M.; Beguería, S.; López-Moreno, J.I. A multiscalar drought index sensitive to global warming: The standardized precipitation evapotranspiration index. *J. Clim.* **2010**, *23*, 1696–1718. [CrossRef]
36. Alam, N.M.; Sharma, G.C.; Moreira, E.; Jana, C.; Mishra, P.K.; Sharma, N.K.; Mandal, D. Evaluation of drought using SPEI drought class transitions and loglinear models for different agro-ecological regions of India. *Phys. Chem. Earth Parts A/B/C* **2017**, *100*, 31–43. [CrossRef]
37. Aryal, S.K.; Zhang, Y.; Chiew, F. Enhanced low flow prediction for water and environmental management. *J. Hydrol.* **2020**, *584*, 124658. [CrossRef]
38. Nathan, R.J.; McMahon, T.A. Evaluation of automated techniques for base flow and recession analyses. *Water Resour. Res.* **1990**, *26*, 465–473. [CrossRef]
39. Cheng, L.; Zhang, L.; Brutsaert, W. Automated selection of pure base flows from regular daily streamflow data: Objective algorithm. *J. Hydrol. Eng.* **2016**, *21*, 6016008. [CrossRef]
40. Wang, L.; Zhang, X.; Wang, S.; Salahou, M.K.; Fang, Y. Analysis and application of drought characteristics based on theory of runs and copulas in Yunnan, Southwest China. *Int. J. Environ. Res. Public Health* **2020**, *17*, 654. [CrossRef]
41. Eckert, C.; Hohberger, J. Addressing endogeneity without instrumental variables: An evaluation of the gaussian copula approach for management research. *J. Manag.* **2023**, *49*, 460–495. [CrossRef]

42. Lin, L.; Zhou, X.; Gao, T.; Zhu, Z.; Qing, Y.; Liao, W.; Lin, W. Herb pairs containing Curcumae Rhizoma (Ezhu): A review of bio-active constituents, compatibility effects and t-copula function analysis. *J. Ethnopharmacol.* **2023**, *319*, 17199. [CrossRef] [PubMed]
43. Poonia, N.; Azad, S. A New Statistical Distribution Derived from a Clayton Copula for Modeling Bivariate Processes. *J. Hydrometeorol.* **2023**, *4*, 659–678. [CrossRef]
44. Huang, S.; Li, Q.; Shu, Z.; Chan, P.W. Copula-based joint distribution analysis of wind speed and wind direction: Wind energy development for Hong Kong. *Wind Energy* **2023**, *6*, 900–922. [CrossRef]
45. Dewick, P.R.; Liu, S. Copula modelling to analyse financial data. *J. Risk Financ. Manag.* **2022**, *5*, 104. [CrossRef]
46. Wang, J.; Rong, G.; Li, K.; Zhang, J. Analysis of Drought Characteristics in Northern Shaanxi Based on Copula Function. *Water* **2021**, *13*, 1445. [CrossRef]
47. Won, J.; Choi, J.; Lee, O.; Kim, S. Copula-based Joint Drought Index using SPI and EDDI and its application to climate change. *Sci. Total Environ.* **2020**, *744*, 140701. [CrossRef] [PubMed]
48. Huang, K.; Fan, Y.R. Parameter uncertainty and sensitivity evaluation of copula-based multivariate hydroclimatic risk assessment. *J. Environ. Inform.* **2021**, *2*, 38. [CrossRef]
49. Salimi, H.; Asadi, E.; Darbandi, S. Meteorological and hydrological drought monitoring using several drought indices. *Appl. Water Sci.* **2021**, *11*, 11. [CrossRef]
50. Chen, H.; Teegavarapu, R.S.V. Spatial and temporal variabilities in baseflow characteristics across the continental USA. *Theor. Appl. Climatol.* **2021**, *143*, 1615–1629. [CrossRef]
51. Breiman, L. Random forests. *Mach. Learn.* **2001**, *45*, 5–32. [CrossRef]
52. Dikshit, A.; Pradhan, B.; Alamri, A.M. Short-Term Spatio-Temporal Drought Forecasting Using Random Forests Model at New South Wales, Australia. *Appl. Sci.* **2020**, *10*, 4254. [CrossRef]
53. Ruiz-Álvarez, M.; Gomariz-Castillo, F.; Alonso-Sarría, F. Evapotranspiration Response to Climate Change in Semi-Arid Areas: Using Random Forest as Multi-Model Ensemble Method. *Water* **2021**, *13*, 222. [CrossRef]
54. Zarei, A.R.; Mahmoudi, M.R.; Moghimi, M.M. Determining the most appropriate drought index using the random forest algorithm with an emphasis on agricultural drought. *Nat. Hazards* **2023**, *115*, 923–946. [CrossRef]
55. Garreau, D.; Luxburg, U. Explaining the explainer: A first theoretical analysis of LIME. In Proceedings of the International Conference on Artificial Intelligence and Statistics, Virtually, 26–28 August 2020; pp. 1287–1296.
56. Jebli, I.; Belouadha, F.Z.; Kabbaj, M.I.; Tilioua, A. Prediction of solar energy guided by pearson correlation using machine learning. *Energy* **2021**, *224*, 120109. [CrossRef]
57. Chen, H.; Huang, S.; Xu, Y.P.; Teegavarapu, R.S.; Guo, Y.; Nie, H.; Xie, H.; Zhang, L. River ecological flow early warning forecasting using baseflow separation and machine learning in the Jiaojiang River Basin, Southeast China. *Sci. Total Environ.* **2023**, *882*, 163571. [CrossRef] [PubMed]
58. Bageshree, K.; Abhishek; Kinouchi, T. A Multivariate Drought Index for Seasonal Agriculture Drought Classification in Semiarid Regions. *Remote Sens.* **2022**, *14*, 3891. [CrossRef]
59. Guo, Y.; Huang, S.; Huang, Q.; Wang, H.; Wang, L.; Fang, W. Copulas-based bivariate socio-economic drought dynamic risk assessment in a changing environment. *J. Hydrol.* **2019**, *575*, 1052–1064. [CrossRef]
60. Herrera-Estrada, J.E.; Satoh, Y.; Sheffield, J. Spatiotemporal dynamics of global drought. *Geophys. Res. Lett.* **2017**, *44*, 2254–2263. [CrossRef]
61. Hao, Z.; AghaKouchak, A. Multivariate standardized drought index: A parametric multi-index model. *Adv. Water Resour.* **2013**, *57*, 12–18. [CrossRef]

water

MDPI

Article

# Optimization of Multi-Reservoir Flood Control Operating Rules: A Case Study for the Chaobai River Basin in China

**Wenhua Wan [1], Yueyi Liu [1,*], Hang Zheng [1], Jianshi Zhao [2], Fei Zhao [3] and Yajing Lu [3]**

1 School of Environment and Civil Engineering, Dongguan University of Technology, Dongguan 523808, China; meviolet@126.com (W.W.); zhenghang00@163.com (H.Z.)
2 State Key Laboratory of Hydro-Science and Engineering, Department of Hydraulic Engineering, Tsinghua University, Beijing 100084, China; zhaojianshi@tsinghua.edu.cn
3 Beijing Water Science and Technology Institute, Beijing 100048, China; zf@bwsti.com (F.Z.); lyj@bwsti.com (Y.L.)
* Correspondence: liuyysn@163.com

**Abstract:** Reservoirs are susceptible to interference from inter-basin water transfer projects intended to relieve serious water shortages. The Central Route of the South-to-North Water Division Project in China has altered the hydrological conditions and water storage status of the terminal reservoir, the Miyun Reservoir, thereby affecting the flood control reliability in the Chaobai River Basin. In this study, a dual-objective five-reservoir operation model was developed, in which reservoir releases are obtained through piecewise linear operating rules. The model considers the flooding risks both downstream of the basin and in the Miyun reservoir area. A parameterization-simulation-optimization approach was employed to obtain the Pareto-optimal front, providing decision-makers with a list of optimal rule parameters to select and match their own risk preferences. All optimized rules could ensure safe operation during the designed floods to be expected once (or more than once) every thousand years. In contrast, the current flood operation schemes largely ignore the water transfer between basins but primarily concentrate on storing water from floods. Thus, the Miyun Reservoir, whose design return period is 1000 years, can easily become filled during a 100-year flood, impeding the system's flood control capacity. Compared to the operating rule optimized in this study, the current schemes result in a 10.5% higher upstream inundation loss and an unsatisfactory 17 million CNY of equivalent water transfer loss.

**Keywords:** multi-reservoir flood control; multi-objective optimization; flood operating rule; Central Route of the South-to-North Water Diversion Project

## 1. Introduction

Floods are natural disasters that pose significant threats to ecology, the environment, infrastructure, agriculture, and even human life. As a representative flood control infrastructure, reservoirs retain floodwater and flatten out the peak flow by intercepting heavy rainfall in flood storage and releasing it after the rainstorm. With frequent water shortages across the world, there is a growing need to operate multiple and multipurpose reservoirs such that the maximum possible water is conserved while ensuring dam safety and flood moderation [1].

Reservoir flood operation is a typical multi-objective problem, which needs to balance the conflicts of the benefits and flood damage [2,3], or flooding risks of multiple sites and different flood stages [4–6]. For multiple reservoir systems, it is more challenging to prioritize the flooding risk targets between the tributaries and mainstream [7], and to allocate the flood-storage capacity among reservoirs [8,9]. These characteristics pose significant modeling and optimization challenges in the flood operation of multi-reservoir systems. Single reservoir flood control research dates back to the 1936 Flood Control Act in the U.S. [10], and Windsor [11] pioneered the use of dynamic programming to optimize operation during flood periods. Multi-reservoir flood control has become a research hot topic due to the increasing water resource demand and the rapid construction of numerous dams in the past two decades, especially in developing countries [12–14]. The major objective of single reservoir flood control is to minimize the peak water level of the dam or the peak flow at the downstream flood control station [6,15]. Multi-reservoir systems have more complex objectives, aiming to minimize flooding risks at multiple flood control stations while also maximizing flood utilization benefits. For example, Moridi and Yazdi [16] employed an improved mixed integer linear programming to determine the optimal allocation of reservoir flood control capacity in the Karkheh multi-reservoir system. Their optimization objectives are reducing both downstream flooding damage and system hydropower loss. Lu et al. [9] applied the theory of large-scale system decomposition and coordination to balance flood control and power generation objectives for the mixed Pi River Basin. Glavan et al. [3] optimized the design of detention reservoirs using scenario-based economic analysis in the Lower Savinja Valley to ensure flood safety while preventing an unacceptable loss of crop yield.

Among the considerable attempts made to derive optimal operating policies for reservoir flood control, it usually can be achieved through implicit stochastic optimization, explicit stochastic optimization, real-time control with forecasting, and a parameterization-simulation-optimization model [12,17,18]. Each of these has its own advantages and disadvantages. Implicit stochastic optimization, also known as deterministic optimization, requires perfect foreknowledge of a flood over a lengthy time horizon [16,19], which, evidently, is not yet reached in the current flood forecasting technologies [20]. Without the presumption of perfect forecasting, explicit stochastic optimization and real-time operation using uncertainty forecast are usually applicable to single reservoir flood control [6,15,21]. Nevertheless, their application to multi-reservoir systems is computationally challenging unless operating rules can be parametrized in some way, as seen in the work of Lu et al. [9]. Parameterization-simulation-optimization involves optimizing the form of operating rules by defining them with a small number of parameters. This method is commonly used in conjunction with intelligent algorithms and widely applied in deriving reservoir scheduling rules for various purposes, including water supply [22,23], hydropower generation [24,25], and flood control [6,7]. These rules are often predefined in the form of rule curves or tables and deduced through fitting methods from optimal processes. This may involve linear regression analysis or surface fitting [18,26] or an iterative simulation-based optimization by adjusting the operating rule parameters [7,27,28]. Overall, studies on flood control for mixed-cascade reservoir systems are still rare.

The operation of reservoir systems faces abrupt hydrological changes caused by climate change and extensive human activities, such as water division. The Miyun Reservoir is the largest reservoir in North China and was the main surface water source for Beijing, the capital of China. The historical continuous decline of streamflow in the Miyun Reservoir has achieved extensive research in terms of causes, potential impacts, and future projections [29–32]. In recent years, climate change has led to an opposite increasing trend of rainfall in northern China [33], leading to a sharp increase in flood control pressure. Additionally, the water division from the Central Route of China's South-to-North Water Transfer Project into Miyun Reservoir has increased its storage, presenting new requirements for flood control scheduling. However, the current reservoir in operation under the influence of inter-basin water transfer predominantly focuses on water supply issues [34,35]. The impact of inter-basin water transfer on flood control situations and thus flood control rule adjustments remain lacking. The Chaobai River Basin, with the Miyun Reservoir as its major controlling infrastructure, was chosen as a case study to demonstrate the optimization of multi-reservoir flood control operations.

The rest of this paper is organized as follows. Section 2 provides detailed information about the study area and the formation of operating rules and optimization models for the mixed five-reservoir system. Section 3 describes the impact of SNWDP on the Miyun Reservoir and the optimization results. The conclusions are drawn in Section 4.

## 2. Methods and Materials

### 2.1. Chaobai River Basin

The study area is the Chaobai River Basin (39°46′–41°49′ N, 115°25′–117°35′ E) in northern China (Figure 1a). The watershed covers an area of 19,400 km$^2$ and has a diverse topography, with higher elevations in the northern region and a gradual decrease in elevation towards the southeast. It is characterized by two major tributaries, namely the Chao River and the Bai River, which merge at the Miyun Reservoir and form the Chaobai River downstream of the reservoir. The annual precipitation and runoff are around 553 mm (range 500–700) and 1653 million m$^3$ (range 615–4320) [36]. The major flood season is June–August. Over 70% of the annual rainfall occurs during this period, predominantly in the form of intense and localized downpours [37].

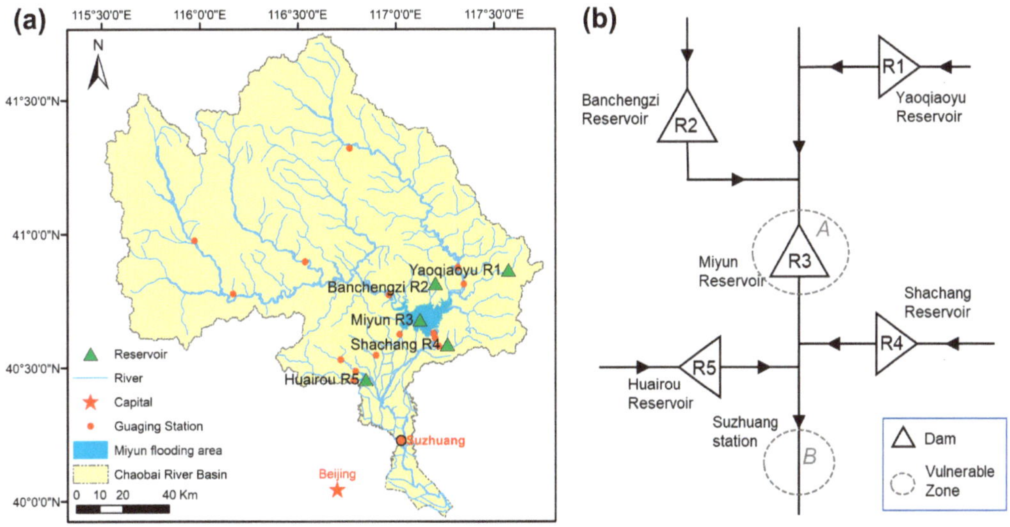


**Figure 1.** Location of the Chaobai River Basin (**a**) and the schematic diagram of the mixed five-reservoir flood control system (**b**).

The flood control system in the Chaobai River Basin consists of a cascade of five reservoirs: the Yaoqiaoyu Reservoir (R1), Banchengzi Reservoir (R2), Miyun Reservoir (R3), Shachang Reservoir (R4), and Huairou Reservoir (R5). The layout is depicted in Figure 1b and numbered from upstream to downstream. The respective flood control characteristic parameters are listed in Table 1. The reservoirs are designed with different seismic intensity levels: R1 and R2 are at 7 degree and the downstream R3, R4, and R5 have a higher design seismic intensity level of 8 degree. Among these reservoirs, the Miyun Reservoir serves as the pivotal control project along the main stream of the basin, with a total storage capacity of 4375 million m$^3$ [38] and a drainage area of approximately 90% of the basin's total area. The Miyun Reservoir has long been the sole surface water resource for Beijing through the Jingmi Canal. Since the introduction of water transfer from the Central Route of the SNWDP in 2015, the reservoir's role shifted to serving as an emergency strategic water resource reserve.

**Table 1.** Characteristic parameters of the Chaobai River flood control system. In the last three rows, the characteristic water level [$m^3$], and in parentheses the respective storage [$10^6$ $m^3$], for each reservoir are given.

| Characteristic Water Level [$m^3$] | Yaoqiaoyu Reservoir (R1) | Banchengzi Reservoir (R2) | Miyun Reservoir (R3) | Shachang Reservoir (R4) | Huairou Reservoir (R5) |
|---|---|---|---|---|---|
| Designed flood control standard [yr] | 100 | 100 | 1000 | 50 | 100 |
| Flood limited water level | 463 (12.1) | 255 (5.75) | 152 (3037) | 165.5 (15.65) | 58 (39.4) |
| Design flood level | 468.1 (17.37) | 258.5 (8.05) | 157.5 (3964) | 167.95 (19.05) | 64.16 (98.2) |
| Check flood level | 469.78 (19.34) | 259.3 (8.63) | 158.5 (4145.4) | 170 (21.2) | 67.73 (144) |

### 2.2. Multi-Reservoir Flood Control Operation Model

The general goal for reservoir flood control is to ensure the safety of both the reservoir itself and the downstream flood control stations. In general, the model objectives during the floods are (1) to minimize the vulnerability (or economic losses) caused by flooding of the reservoir in terms of the reservoir risk water level and the duration of such a high level, and (2) to minimize the damage due to peak outflow from the reservoir at critical locations downstream of the reservoir.

Flood control in the study area is unique because both the upstream reservoir area and downstream flood space have undergone significant urbanization, and the urban development within the Beijing metropolitan area has placed extremely high demands on flood safety in Miyun Reservoir. The flood control objects in this study are (1) the upstream Miyun Reservoir area (Zone A in Figure 1) and (2) the downstream Suzhuang flood control station (Zone B).

The potential inundation losses in the Miyun Reservoir area are due to the continuous decline in the reservoir water level over the past few decades [30]. The long-term exposed reservoir's original inundation area has been developed and inhabited by humans. The flooding risk of the area upstream of the Miyun Reservoir is simulated using the economic losses of human settlements, infrastructure, and agriculture. Forested areas will be submerged when the water level is between 152 m (the flood limited water level, FLWL) and 155 m. When the water level exceeds 155 m, seven towns will be submerged. When the water level reaches 158.5 m (the check flood level, CFL), the water supply security will be impaired. The standardized inundation losses at different water levels under the CFL are collected and calculated (Figure 2a). Therefore, flooding damage can occur at any time when the water level is higher than the FLWL, which is unique and unlike others that are a threat only when the risky water level is exceeded.

The downstream area of the Chaobai River Basin has important towns and a major transport railway and highway. The Suzhuang station is designated as the flood control site. The discharge is jointly managed by the Miyun (R3), Shachang (R4), and Huairou (R5) reservoirs. The safe carrying capacity at Suzhuang station is 600 $m^3/s$, exceeding which would result in varying degrees of flooding. Through flood routing simulations, the extent of inundation and damage in the channel can be determined. Figure 2b illustrates the relationship between the total discharge and the downstream losses due to flooding.

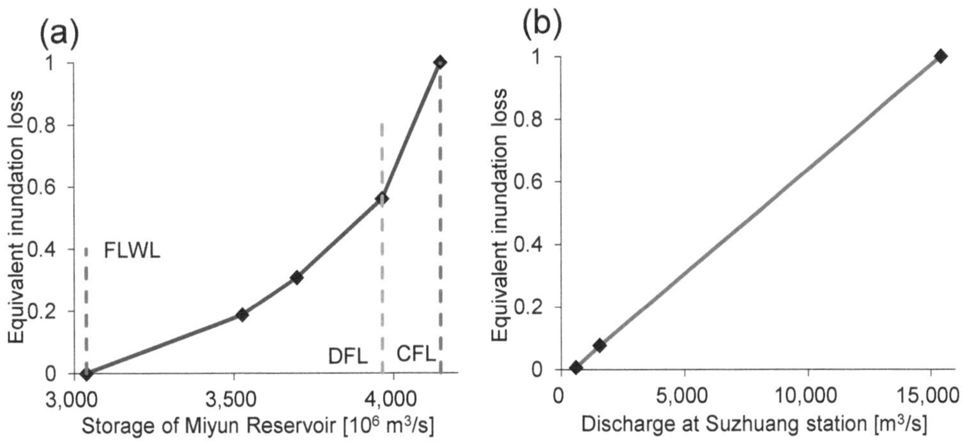


**Figure 2.** Relationships between inundation loss in the Miyun Reservoir area and the reservoir storage (**a**), and the downstream losses and discharge in the Suzhuang flood control section (**b**).

The inundation losses in the Chaobai River Basin are semi-quadratic (or linear) functions of the decision variables. To simplify the model calculation and ensure the stability of the optimization process across time periods, we use the quadratic function instead of the flooding damage as the target and normalized it as follows:

$$\text{Minimize}F_1 = \sum_{t=1}^{T}\left(\frac{V_3^t - V_{FLWL,3}}{V_{DFL,3} - V_{FLWL,3}}\right)^2 \tag{1}$$

$$\text{Minimize}F_2 = \sum_{t=1}^{T}\left(\frac{R_3^t + R_4^t + R_5^t}{R_{max,3} + R_{max,4} + R_{max,5}}\right)^2 \tag{2}$$

These objectives are subjective to the following constraints.
Water balance constraint

$$V_i^t = \begin{cases} V_i^{t-1} + I_i^t\Delta t - R_i^t\Delta t & i = 1,2,4,5 \\ V_i^{t-1} + \left(I_i^t + R_1^t + R_2^t - R_i^t\right)\Delta t & i = 3 \end{cases} \tag{3}$$

Reservoir water storage or level constraint

$$Z_{min,i} \le Z\left(V_i^t\right) \le Z_{max,i} \tag{4}$$

Outflow constraint

$$R_{min,i} \le R_i^t \le R_{max,i} \tag{5}$$

where, $V_i^{t-1}$ and $V_i^t$ are the beginning and ending storages of reservoir $i \in \{1,2,\ldots 5\}$ in period $t$ ($m^3$), respectively. Specifically, $V_3^t$ is the ending storage of the Miyun Reservoir. $V_{FLWL,3}$ and $V_{DFL,3}$ are the storages corresponding to the flood limited water level and designed flood level ($m^3$), respectively. $I$, $R$, $Z$ are the natural inflow ($m^3/s$), reservoir outflow ($m^3/s$), and reservoir water level (m), respectively. The intermediate flow between reservoirs, i.e., regions between the reservoir and the flood control section, is ignored. The Miyun Reservoir (R3) receives the discharges from the Yaoqiaoyu Reservoir (R1) and the Banchengzi Reservoir (R2). $R_3^t$, $R_4^t$, and $R_5^t$ are the discharges of the Miyun (R3), Shachang (R4), and Huairou (R5) reservoirs ($m^3/s$), the sum of which is the flow rate at the downstream Suzhuang flood control section. $Z_{min,i}$ and $Z_{max,i}$ are the lower and upper limits of the reservoir water level (m), respectively, and are typically the dead water level and check flood level (CFL), respectively, during the flood season. $R_{min,i}$ and $R_{max,i}$ are the boundaries of the outflow constraints and are equal to the ecological flow requirement and the reservoir spillway capacity of the respective reservoir at the check flood level ($m^3/s$).

### 2.3. Operating Rules

Flood control operating rules are used to make release decisions based on the current hydrological conditions, such as the reservoir water level and incoming flow rates, thereby achieving conservative safety operation during a flood. These rules can be established using empirical relationships, simulation models, and optimization methods. The release schemes are usually presented in the form of tables or mathematical functions, with

$$Q_{out} = f(Q_{in}, Z) = \begin{cases} Q_{out,1} & Q_{in} \in (Q_{min,1}, Q_{max,1}),\ Z \in (Z_{min,1}, Z_{max,1}) \\ Q_{out,2} & Q_{in} \in (Q_{min,2}, Q_{max,2}),\ Z \in (Z_{min,2}, Z_{max,2}) \\ & \vdots \\ Q_{max,K} & Q_{in} \in (Q_{min,K}, Q_{max,K}),\ Z \in (Z_{min,K}, Z_{max,K}) \end{cases} \quad (6)$$

where $Q_{out}$ (i.e., $R_i^t$ in Equations (2) and (5)) is the reservoir release to be decided ($m^3/s$), which is usually a piecewise parametric function of the reservoir states, i.e., the inflow $Q_{in}$ ($m^3/s$) and water level $Z$ (m). $Q_{min,k}$, $Q_{max,k}$, $Z_{min,k}$, and $Z_{max,k}$, $k$ = 1,2,. . .$K$, are the hierarchical boundaries of the classified reservoir inflow and water level.

In general, reservoirs follow uniform rules when confronted with frequent flood events. However, a uniform parameterization scheme cannot adequately address the specific flood safety requirements for the Chaobai River Basin. Distinct parameters have been proposed for a similar set of rules for each reservoir, to cope with floods with different frequencies. We use the following widely used linear decision rules [7,27,39] with hourly regulation and four-lever hierarchy:

$$R_i^t = \begin{cases} 0 & V_i^t < V_{FLWL,i} \\ \min(\alpha_i Q_{in,i}^t + \beta_i \frac{(V_i^{t-1} - V_{FLWL,i})}{\Delta t}, R_{max,i}) & V_{FLWL,i} \le V_i^t < V_{DFL,i} \\ R_{Dmax,i} & V_{DFL,i} \le V_i^t < V_{CFL,i} \\ R_{max,i} & V_i^t \ge V_{CFL,i} \end{cases} \quad (7)$$

In the case of extreme floods with a return period of over 100 years (e.g., 1000-year-flood), the form of the rule needs to be slightly modified for reservoirs other than the Miyun Reservoir. The discharge increases with $i = 1, 2, 4, 5$:

$$R_i^t = \begin{cases} R_{Fmax,i} & V_i^t < V_{FLWL,i} \\ \min(\alpha_i Q_{in,i}^t + \beta_i \frac{(V_i^{t-1} - V_{FLWL,i})}{\Delta t} + \gamma_i I_{max,i}, R_{max,i}) & V_{FLWL,i} \le V_i^t < V_{DFL,i} \\ R_{Dmax,i} & V_{DFL,i} \le V_i^t < V_{CFL,i} \\ R_{max,i} & V_i^t \ge V_{CFL,i} \end{cases} \quad (8)$$

where the reservoir release is decided as a linear function of both the reservoir inflow and the water level. $Q_{in,i}^t$ is the reservoir inflow from the flood hydrograph and upstream releases ($m^3/s$); $R_{Fmax,i}$, $R_{Dmax,i}$, and $R_{max,i}$ are the maximum allowable discharges ($m^3/s$) under FLWL, DFL, and CFL, respectively. The parameters $\alpha_{i=1,2\ldots5}$, $\beta_{i=1,2\ldots5}$, and $\gamma_{i=4,5}$ are the decision variables that need to be optimized in this study.

### 2.4. Investigated Floods

The 100-year design flood hydrographs for each reservoir area are illustrated in Figure 3. Table 2 lists the peak flood discharges for each reservoir under different return periods. By proportionally scaling the representative floods using the peak flows and 3-day flood quantities (not shown), the 72-h design flood hydrographs for different frequencies are obtained.

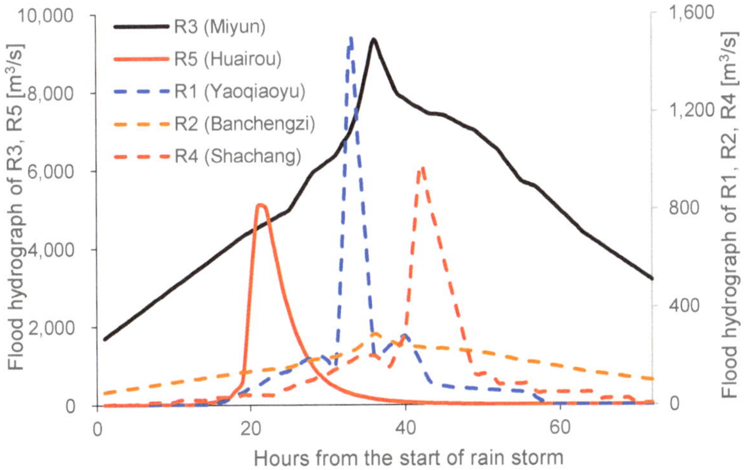


**Figure 3.** 72-h designed flood hydrographs for the designed 100-year-flood.

**Table 2.** Low-recurrence-interval design flood peak flow discharge.

| Flood return period [yr] | 1000 | 100 | 50 | 20 | 10 |
|---|---|---|---|---|---|
| Frequency [%] | 0.1 | 1 | 2 | 5 | 10 |
| | | | Peak flow [m$^3$/s] | | |
| R1 (Yaoqiaoyu) | 2290 | 1500 | 1280 | 980 | 763 |
| R2 (Banchengzi) | 457 | 288 | 195 | 78 | 67.2 |
| R3 (Miyun) | 15,800 | 9320 | 7460 | 5120 | 3480 |
| R4 (Shachang) | 1510 | 975 | 800 | 590 | 419 |
| R5 (Huairou) | 7710 | 5059 | 4270 | 3280 | 2440 |

### 2.5. NSGA-II Solving Method

Due to the non-differentiable nature of the predefined operating rules (Section 2.3) and the nonlinear relationship between rule parameters and flood damage, classical optimization methods like linear programming are not suitable for optimizing the model. Instead, a simulation-based optimization approach is employed to derive the rule parameters.

In this paper, the non-dominated sorting genetic algorithm II algorithm (NSGA-II) [40] is implemented to identify the large set of Pareto solutions to this simulation-optimization multi-objective model (Section 2.2). The NSGA-II eliminates the need for explicit coordination among multiple objectives and enhances the basic genetic algorithm by incorporating fast non-dominated sorting and crowding distance mechanisms. An elite preservation strategy in NSGA-II ensures diverse solutions that effectively approach the Pareto optimal front in a wide and uniform manner. Hojjati et al. [41] have found that NSGA-II provides better approximations of the true Pareto optimal surface, depicting tradeoffs between objectives, compared to multi-objective particle swarm optimization. The NSGA-II algorithm is well-established and has been demonstrated to have strong optimization capabilities in both theoretical test functions and practical production problems [25,40]. The obtained Pareto front provides valuable insights into available compromising strategies for decision-makers.

The operating rules (Section 2.3) are optimized for various representative flood frequencies, starting from the FLWL of each reservoir. For each representative flood (Section 2.4), a set of initial decision variables (i.e., the rule parameters) is randomly generated. Simulation methods are then used to sequentially update the reservoir states. The operation process and objective function values are evaluated (Section 2.2). Optimized using the

NSGA-II algorithm, all non-dominated optimal solutions, including the Pareto front and corresponding rule parameters, are obtained.

## 3. Results and Discussion

### 3.1. Impact of Inter-Basin Water Transfer on the Flood Control Situation of the Miyun Reservoir

Figure 4 illustrates the historical annual end-of-year storage of the Miyun Reservoir. Before 1995, the reservoir's water level was close to the FLWL, which has undergone multiple adjustments [39] and currently stands at 152 m (3037 million $m^3$). Since then, a combination of factors including reduced rainfall, has led to consistently low reservoir storage (Figure 4). Flood control was not a major consideration during the flood season. Instead, the reservoir intercepted all incoming floodwater for non-flood season water supply.

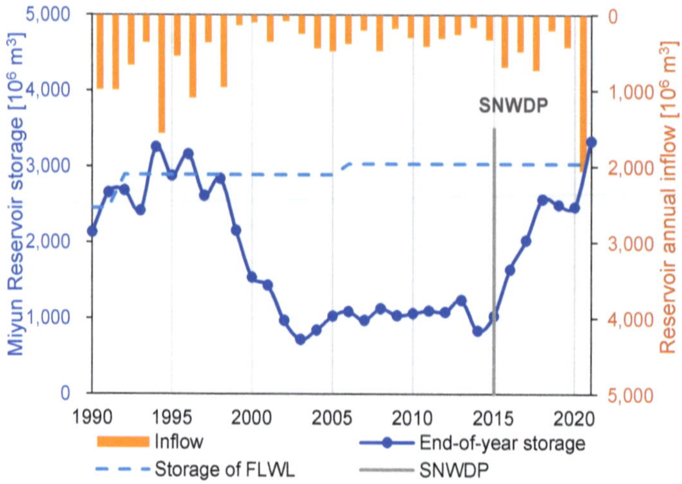


**Figure 4.** Annual available water resources into the Miyun Reservoir (including external transfer), end-of-year reservoir storage, and the historical flood limited water level (FLWL) from 1990 to 2021. The Central Route of the South-to-North Water Diversion Project (SNWDP) commenced water transfer in 2015.

The Miyun reservoir began to receive water from the Central Route of the SNWDP in 2015, accounting for approximately 10–30% of the total inflow to the reservoir. Along with the decreasing water supply and increasing trend of extreme floods recently, water storage in the Miyun Reservoir has been steadily rising and reached its peak in 2021 (Figure 4). During the flood season in 2021, the basin experienced 63% more precipitation than the average, and the historically largest flood event, which occurred from 30 July to 2 August. With these changes, the reservoir now seeks to maximize water storage due to the high cost of inter-basin water transfer while mitigating potential losses caused by heavy rainfall. The discharge during flood regulation consists of a combination of upstream floodwater and water transferred from southern China. Consequently, the equivalent economic losses of the Miyun Reservoir discharge can be evaluated as the product of the discharge volume and the unit price of the transferred water.

### 3.2. Flood Regulation

Taking the 20-year flood (within design standard) and 1000-year flood (beyond design standard) events as examples, Figure 5 shows the Pareto front of the dual-objective model using the NSGA-II method. Objective F1 aims to minimize the water level of the Miyun Reservoir, thus reducing the losses due to inundation in the reservoir area (Equation (1)).

Objective F2 focuses on controlling the system downstream discharge in the Suzhuang section, thus constraining the flooding risk in the downstream river channel (Equation (2)). These objectives are in competition and reducing one will enlarge the other. Under the two different frequency floods, the upstream target values are in a similar range, but the downstream target values are six times larger for the 1000-year flood than for the 20-year flood (Figure 5).

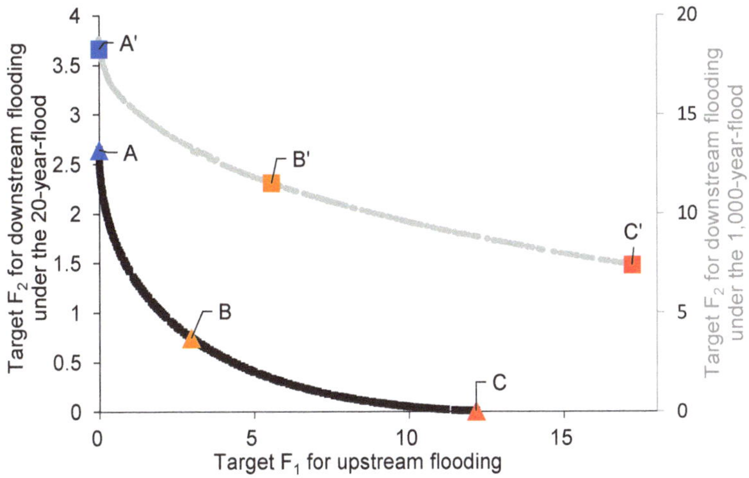


**Figure 5.** The Pareto front for the dual-objective operation model under a 20-year flood and a 1000-year flood. The six points (A, B, C, A′, B′, and C′) represent the six solutions under different flood conditions and distinct objective prioritization.

Each point on the Pareto front represents an optimal solutions alternative obtained through the optimization of the reservoir operation. Figures 5 and 6a,b show a comparison of the reservoir operation process under the 20-year flood. The blue point (line) labeled A on the Pareto front indicates full focus on minimizing upstream inundation (objective $F_1$). In this case, the water level of the Miyun Reservoir remains relatively stable at the FLWL (Figure 6a). To achieve this, almost all inflow is discharged, leading to high flow rates in the Suzhuang station, and the peak value even surpasses the system's natural inflow (Figure 6b). As the concern shifts toward prioritizing the downstream flood safety (the trade-off point B on the Pareto front), the reservoir storage gradually increases (orange curve in Figure 6a) from the FLWL (3037 million $m^3$) to 154.7 m (3422 million $m^3$) within 72 h. In return, the system release exhibits a noticeable peak attenuation effect (Figure 6b). The goal of red point (curve) C is to minimize downstream flooding by intercepting as much incoming floodwater as possible in the Miyun Reservoir.

Under the 1000-year flood, the system is still capable of safeguarding the entire basin (Figure 6c,d). None of the selected three scenarios causes the water level of the Miyun Reservoir to surpass the DFL. As the system's designed flood has two peaks, the reservoirs effectively capture the first peak while the attenuation effect on the second peak is noticeably diminished. This weakness can be attributed to the shrinking flood control storage of the reservoirs during the operation.

The optimal decision variables (i.e., rule parameters) under each frequency design flood are provided in the Supplementary Materials (Table S1). The rule-based operation typically involves two steps: (1) evaluation of the magnitude of the incoming flood by comparing the forecast peak flow and quantity to the historical flood records; and (2) releasing the water according to the operating rules with the corresponding parameters and Equations (6)–(8).

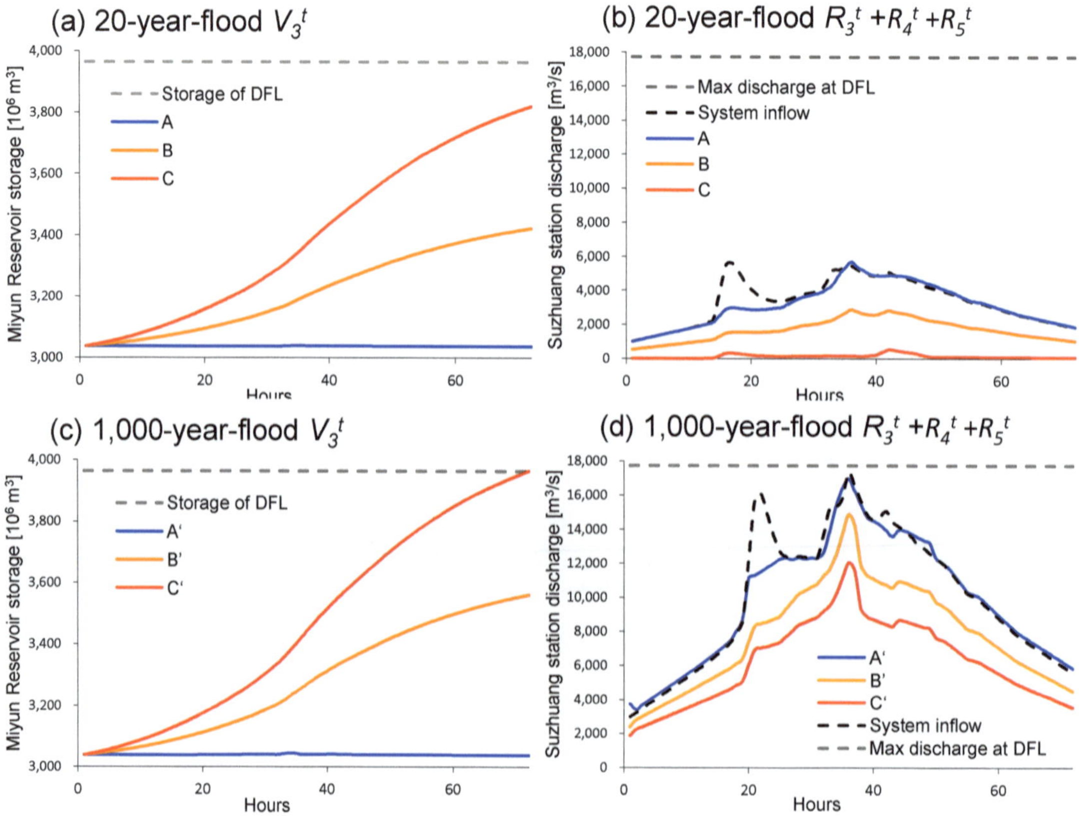


**Figure 6.** The reservoir operation process in terms of the Miyun Reservoir storage (**a**,**c**) and the basin's downstream total discharge (**b**,**d**) corresponding to the points A, B, C, A′, B′, and C′ in Figure 5.

### 3.3. Compariosn to the Current Operating Rules

The existing flood control rules for the five reservoirs listed in Table 3 are designed independently for the single-objective single-reservoir condition. In general, the reservoir discharge varies in different characteristic water level ranges. When reservoir storage is below the FLWL, all incoming floodwater is stored in the reservoirs. When water level exceeds FLWL, reservoirs with small flood control capacity, i.e., R1, R2, and R4, are designed to discharge either up to the maximum capacity or at predetermined flow rates. The Miyun Reservoir (R3) and Huairou Reservoir (R5) adjust their discharge considering the inflow rate more extensively, with small release under low inflow conditions and high release when there is a significant inflow.

Taking the 100-year flood as an example, the current rules lead to greater flood control loss in both the upstream and downstream areas compared to the dual-objective multi-reservoir integrated optimization result (Figure 7a). For illustrative purposes, we compare the current rules with the closest optimal Pareto solution, the rule parameters of which are listed in Table 3.

**Table 3.** The current flood control operating rules for individual reservoirs and the optimized rule parameters for the integrated operation of the five reservoirs on the Chaobai River when confronting a 100-year flood.

| Criteria | R1 | R2 | R3 | R4 | R5 | Upstream Damage | Downstream Damage | Release from R3 |
|---|---|---|---|---|---|---|---|---|
| Release according to the current rules | | | | | | [economic equivalence] | | $[10^6\ m^3]$ |
| Z < FLWL | 0 | 0 | 0 | 0 | 0 | 15.91 | 9.1 | 479 |
| FLWL ≤ Z < DFL | $R_{Fmax,1}$ | 80 | {600, 1000, 1500} [a] | $Q^t_{in,4}$ | $\frac{Q^t_{in,5} R_{Dmax,5}}{5059}$ [a] | | | |
| DFL ≤ Z < CFL | $R_{Dmax,1}$ | 200 | $\{Q^t_{in,3}, R_{Dmax,3}\}$ [a] | {420, 670} [a] | $R_{Dmax,5}$ | | | |
| Z ≥ CFL | $R_{max,1}$ | $R_{max,2}$ | $\{Q^t_{in,3}, R_{max,3}\}$ [a] | $R_{max,4}$ | $R_{max,5}$ | | | |
| Rule parameters for the Pareto optimal | | | | | | | | |
| $\alpha_i$ | 0.2698 | 0.3616 | 0.3528 | 0.4129 | 0.4032 | 14.4 | 8.9 | 471.6 |
| $\beta_i$ | 0.5586 | 0.6210 | 0 | 0.4779 | 0.0114 | | | |

Note: [a]: the specific determination of how much to release is found in the respective reservoir flood operation scheme.

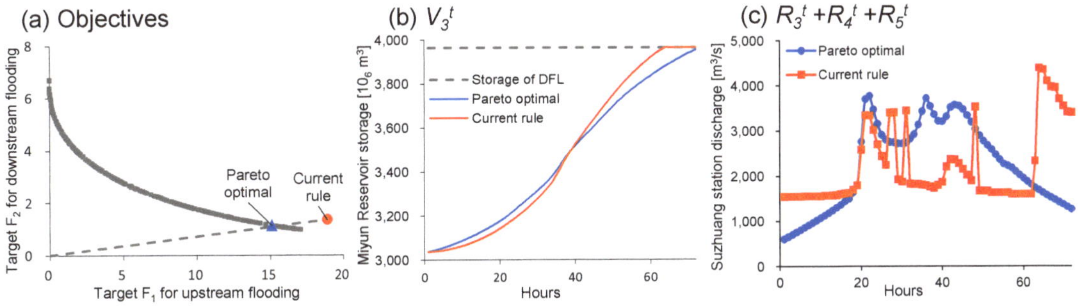


**Figure 7.** Comparison of operation under current rules and optimized rules in terms of the objective values (**a**), the Miyun Reservoir storage process (**b**), and the basin's downstream total discharge (**c**).

Under the current rules, the primary focus of the Miyun Reservoir is on inflow interception, regardless of the background of the SNWDP. The reservoir tends to fill up prematurely, thus reducing its flood control capacity in the latter stages. It is evident from the storage process shown in Figure 7b that the current rules could rapidly increase the water level to the DFL, after which all inflow must be released. Conversely, the optimized rules (blue curve in Figure 7b) allow for water storage at lower inflow rates and dynamically adjust the outflow proportionally to the inflow and storage, resulting in more gradual filling. The optimized rules, with emphasis on reservoir storage, i.e., discharging in response to both incoming water and dynamic changes in storage, lead to a more continuous reservoir outflow process. This helps avoid the frequent abrupt changes in discharge observed under the current rules (Figure 7c).

In terms of flood damage, the current rules for a 100-year flood lead to 10.5% more loss in the upstream reservoir area compared to the optimized rules. Downstream inundation losses are similar, but the Miyun Reservoir incurs an additional 7.4 million $m^3$ spill. Assuming a unit water price of 2.33 CNY for the water transferred to Beijing via the Central Route of the SNWDP [42], the surplus water is worth about around 17 million CNY.

## 4. Conclusions

In this study, a multi-objective model for mixed-reservoir flood control operation was developed. The operating rules were optimized using the NSGA-II algorithm. The mixed five-reservoir system in the Chaobai River Basin was selected because the major flood control reservoir, the Miyun Reservoir, has been significantly influenced in terms of both its water inflow and demand by the Central Route of the SNWDP. Unlike conventional flood control models whose upstream risk is dam overtopping, the Miyun Reservoir faces unique challenges: the upstream reservoir area faces inundation loss as long as the water level exceeds the FLWL and the loss increases non-linearly with the increasing inundation depth. The piecewise linear rules are directly proportional to the inflow and reservoir storage. These linear rules are operational simplicity and in line with the current operation schemes in the Chaobai River Basin. An integrated simulation-optimization frame is adopted for the joint reservoirs. This means that the rules are optimized by accounting for the interplay, such as the compensation effects, among the reservoirs, thereby yielding more reasonable rules than optimizing each reservoir independently.

The model provides a range of rule parameter combinations that accommodate the different preferences of the targets. A comparison with the current flood operating rules revealed that the optimized rules can avoid premature reservoir filling and decrease inundation losses in the Miyun Reservoir area and downstream of the basin.

In this study, different combinations of rule parameters are proposed to address floods of varying frequencies. While these combinations could effectively handle floods of the corresponding magnitudes, the effectiveness of the operating rules in different flood coincidences in the mainstream and tributaries has not been investigated. In addition, only piecewise linear operating rules are considered. To enhance the adaptability and robustness of the system flood control operation, future research should leverage the optimized parameter library derived from designed floods and the actual normal flood events. The performances of other types of operating rules should be compared to determine the most suitable rules for the control of basin-wide floods in future studies. The operating rules and rule parameters could be further adjusted to provide more dynamic decision support for operators under typical flood conditions and incorporating real-time weather forecasts.

**Supplementary Materials:** The following supporting information can be downloaded at: https://www.mdpi.com/article/10.3390/w15152817/s1, Table S1: Parameters for operating rules.

**Author Contributions:** Conceptualization, W.W. and Y.L. (Yueyi Liu); methodology and formal analysis, W.W. and H.Z.; data resources, F.Z. and Y.L. (Yajing Lu); writing—original draft preparation, W.W., H.Z. and J.Z. All authors provided critical feedback and helped shape the manuscript and revision. All authors have read and agreed to the published version of the manuscript.

**Funding:** This study was financially supported by the National Natural Science Foundation of China (U2040206), (52179009) and (51909035).

**Conflicts of Interest:** The authors declare no conflict of interest.

## References

1. Jain, S.K.; Shilpa, L.S.; Rani, D.; Sudheer, K.P. State-of-the-art review: Operation of multi-purpose reservoirs during flood season. *J. Hydrol.* **2023**, *618*, 129165. [CrossRef]
2. Wan, W.; Zhao, J.; Lund, J.R.; Zhao, T.; Lei, X.; Wang, H. Optimal Hedging Rule for Reservoir Refill. *J. Water Res. Plan. Man.* **2016**, *142*, 4016051. [CrossRef]
3. Glavan, M.; Cvejić, R.; Zupanc, V.; Knapič, M.; Pintar, M. Agricultural production and flood control dry detention reservoirs: Example from Lower Savinja Valley, Slovenia. *Environ. Sci. Policy* **2020**, *114*, 394–402. [CrossRef]
4. Ding, W.; Zhang, C.; Cai, X.; Li, Y.; Zhou, H. Multi-objective hedging rules for flood water conservation. *Water Resour. Res.* **2017**, *53*, 1963–1981. [CrossRef]
5. Zhao, T.; Zhao, J.; Lund, J.R.; Yang, D. Optimal Hedging Rules for Reservoir Flood Operation from Forecast Uncertainties. *J. Water Res. Plan. Man.* **2014**, *140*, 4014041. [CrossRef]
6. Chou, F.N.; Wu, C. Stage-wise optimizing operating rules for flood control in a multi-purpose reservoir. *J. Hydrol.* **2015**, *521*, 245–260. [CrossRef]
7. Lei, X.; Zhang, J.; Wang, H.; Wang, M.; Khu, S.; Li, Z. Deriving mixed reservoir operating rules for flood control based on weighted non-dominated sorting genetic algorithm II. *J. Hydrol.* **2018**, *564*, 967–983. [CrossRef]
8. Meng, X.; Chang, J.; Wang, X.; Wang, Y.; Wang, Z. Flood control operation coupled with risk assessment for cascade reservoirs. *J. Hydrol.* **2019**, *572*, 543–555. [CrossRef]

9. Lu, Q.; Zhong, P.; Xu, B.; Zhu, F.; Huang, X.; Wang, H.; Ma, Y. Stochastic programming for floodwater utilization of a complex multi-reservoir system considering risk constraints. *J. Hydrol.* **2021**, *599*, 126388. [CrossRef]
10. Arnold, J.L. *The Evolution of the 1936 Flood Control Act*; Office of History; US Army Corps of Engineers: Fort Belvoir, VA, USA, 1988.
11. Windsor, J.S. Optimization model for the operation of flood control systems. *Water Resour. Res.* **1973**, *9*, 1219–1226. [CrossRef]
12. Labadie, J.W. Optimal operation of multireservoir systems: State-of-the-art review. *J. Water Res. Plan. Man.* **2004**, *130*, 93–111. [CrossRef]
13. Guo, S.; Chen, J.; Li, Y.; Liu, P.; Li, T. Joint operation of the multi-reservoir system of the Three Gorges and the Qingjiang cascade reservoirs. *Energies* **2011**, *4*, 1036–1050. [CrossRef]
14. Rahimi, H.; Ardakani, M.K.; Ahmadian, M.; Tang, X. Multi-reservoir utilization planning to optimize hydropower energy and flood control simultaneously. *Environ. Process.* **2020**, *7*, 41–52. [CrossRef]
15. Qi, Y.; Yu, J.; Li, X.; Wei, Y.; Miao, Q. Reservoir flood control operation using multi-objective evolutionary algorithm with decomposition and preferences. *Appl. Soft Comput.* **2017**, *50*, 21–33. [CrossRef]
16. Moridi, A.; Yazdi, J. Optimal allocation of flood control capacity for multi-reservoir systems using multi-objective optimization approach. *Water Resour. Manag.* **2017**, *31*, 4521–4538. [CrossRef]
17. Ahmad, A.; El-Shafie, A.; Razali, S.F.M.; Mohamad, Z.S. Reservoir optimization in water resources: A review. *Water Resour. Manag.* **2014**, *28*, 3391–3405. [CrossRef]
18. Celeste, A.B.; Billib, M. Evaluation of stochastic reservoir operation optimization models. *Adv. Water Resour.* **2009**, *32*, 1429–1443. [CrossRef]
19. Kim, Y.; Sun, B.; Kim, P.; Jo, M.; Ri, T.; Pak, G. A study on optimal operation of gate-controlled reservoir system for flood control based on PSO algorithm combined with rearrangement method of partial solution groups. *J. Hydrol.* **2021**, *593*, 125783. [CrossRef]
20. Zhao, T.; Cai, X.; Yang, D. Effect of streamflow forecast uncertainty on real-time reservoir operation. *Adv. Water Resour.* **2011**, *34*, 495–504. [CrossRef]
21. Shim, K.; Fontane, D.G.; Labadie, J.W. Spatial decision support system for integrated river basin flood control. *J. Water Res. Plan. Man.* **2002**, *128*, 190–201. [CrossRef]
22. Wan, W.; Zhao, J.; Wang, J. Revisiting water supply rule curves with hedging theory for climate change adaptation. *Sustainability* **2019**, *11*, 1827. [CrossRef]
23. Chen, L.; McPhee, J.; Yeh, W.W.G. A diversified multiobjective GA for optimizing reservoir rule curves. *Adv. Water Resour.* **2007**, *30*, 1082–1093. [CrossRef]
24. Afshar, A.; Shafii, M.; Haddad, O.B. Optimizing multi-reservoir operation rules: An improved HBMO approach. *J. Hydroinform.* **2011**, *13*, 121–139. [CrossRef]
25. Liu, P.; Guo, S.; Xu, X.; Chen, J. Derivation of Aggregation-Based Joint Operating Rule Curves for Cascade Hydropower Reservoirs. *Water Resour. Manag.* **2011**, *25*, 3177–3200. [CrossRef]
26. Zhang, J.; Liu, P.; Wang, H.; Lei, X.; Zhou, Y. A Bayesian model averaging method for the derivation of reservoir operating rules. *J. Hydrol.* **2015**, *528*, 276–285. [CrossRef]
27. Liu, P.; Li, L.; Chen, G.; Rheinheimer, D.E. Parameter uncertainty analysis of reservoir operating rules based on implicit stochastic optimization. *J. Hydrol.* **2014**, *514*, 102–113. [CrossRef]
28. Nourani, V.; Rouzegari, N.; Molajou, A.; Hosseini Baghanam, A. An integrated simulation-optimization framework to optimize the reservoir operation adapted to climate change scenarios. *J. Hydrol.* **2020**, *587*, 125018. [CrossRef]
29. Bao, Z.; Fu, G.; Wang, G.; Jin, J.; He, R.; Yan, X.; Liu, C. Hydrological projection for the Miyun Reservoir basin with the impact of climate change and human activity. *Quatern Int.* **2012**, *282*, 96–103. [CrossRef]
30. Wang, Z. Status and causes of storage change in Miyun Reservoir. *Beijing Water* **2013**, 13–16. (In Chinese)
31. Xu, W.; Zhao, J.; Zhao, T.; Wang, Z. Adaptive Reservoir Operation Model Incorporating Nonstationary Inflow Prediction. *J. Water Res. Plan. Man.* **2014**, *141*, 4014099. [CrossRef]
32. Ma, H.; Yang, D.; Tan, S.K.; Gao, B.; Hu, Q. Impact of climate variability and human activity on streamflow decrease in the Miyun Reservoir catchment. *J. Hydrol.* **2010**, *389*, 317–324. [CrossRef]
33. Yao, J.; Sun, S.; Zhai, H.; Feger, K.; Zhang, L.; Tang, X.; Li, G.; Wang, Q. Dynamic monitoring of the largest reservoir in North China based on multi-source satellite remote sensing from 2013 to 2022: Water area, water level, water storage and water quality. *Ecol. Indic.* **2022**, *144*, 109470. [CrossRef]
34. Wang, Q.; Zhou, H.; Liang, G.; Xu, H. Optimal Operation of Bidirectional Inter-Basin Water Transfer-Supply System. *Water Resour. Manag.* **2015**, *29*, 3037–3054. [CrossRef]
35. Wan, W.; Guo, X.; Lei, X.; Jiang, Y.; Wang, H. A novel optimization method for multi-reservoir operation policy derivation in complex inter-basin water transfer system. *Water Resour. Manag.* **2018**, *32*, 31–51. [CrossRef]
36. Wang, Y.; Xue, X.; Zhang, X. The Drought Assessment and Its Impact Analysis in Chaobai River Basin from 1980 to 2012. *Clim. Chang. Res. Lett.* **2014**, *3*, 52–60. [CrossRef]
37. Zhou, X.; Pan, X.; Zhu, Y.; Hu, Y. Research of the Mean Annual Water Balance during 1980–2013 in Chaobai River Basin. *J. Nat. Resour.* **2016**, *31*, 649–657. (In Chinese) [CrossRef]
38. Hu, Y.; Zhou, L.; Yan, M.; Dong, X.; Wu, Z.; Wei, C. The Valuation and Compensation Policies Suggestions of Main Capitals and Services of Miyun Reservoir. *J. Nat. Resour.* **2007**, *22*, 497–506. (In Chinese)

39. Zhang, M.; Xue, Z.; Pan, L.; Wang, Z. Analysis on dispatching potential of Miyun Reservoir for extreme rainstorm under high water level condition. *Beijing Water* **2023**, 7–13. (In Chinese) [CrossRef]
40. Deb, K.; Pratap, A.; Agarwal, S.; Meyarivan, T. A fast and elitist multiobjective genetic algorithm: NSGA-II. *IEEE Trans. Evol. Comput.* **2002**, *6*, 182–197. [CrossRef]
41. Hojjati, A.; Monadi, M.; Faridhosseini, A.; Mohammadi, M. Application and comparison of NSGA-II and MOPSO in multi-objective optimization of water resources systems. *J. Hydrol. Hydromech.* **2018**, *66*, 323–329. [CrossRef]
42. Crow-Miller, B.; Webber, M. Of maps and eating bitterness: The politics of scaling in China's South-North Water Transfer Project. *Political Geogr.* **2017**, *61*, 19–30. [CrossRef]

water

MDPI

Article

# Spatial and Temporal Assessment of Baseflow Based on Monthly Water Balance Modeling and Baseflow Separation

**Huawei Xie [1,2], Haotian Hu [1,2], Donghui Xie [3], Bingjiao Xu [1,2], Yuting Chen [4], Zhengjie Zhou [5], Feizhen Zhang [6] and Hui Nie [1,2,*]**

1 College of Hydraulic and Environmental Engineering, Zhejiang University of Water Resources and Electric Power, Hangzhou 310018, China; xiehw@zjweu.edu.cn (H.X.); 2021b20037@stu.zjweu.edu.cn (H.H.); 2021b20045@stu.zjweu.edu.cn (B.X.)
2 Key Laboratory for Technology in Rural Water Management of Zhejiang Province, Hangzhou 310018, China
3 Ningbo Yuanshui Company Limited, Ningbo 315000, China
4 Zhejiang Water Conservancy Development Planning Research Center, Hangzhou 310012, China; chenyt@zjwater.gov.cn
5 Zhejiang Institute of Hydraulics & Estuary (Zhejiang Institute of Marine Planning and Design), Hangzhou 310020, China; zzjwater@163.com
6 Hangzhou Nanpai Engineering Construction Management Service Center, Hangzhou 310020, China; hzzfz2005@163.com
* Correspondence: nieh@zjweu.edu.cn

**Abstract:** Baseflow is the part of streamflow that is mainly replenished by groundwater. The protection of the biological environment and the growth of its water resources greatly depend on the spatial and temporal evolution of baseflow. Therefore, the Baizhiao (BZA) and Shaduan (SD) catchments of the Jiaojiang River Basin (JRB) in the Zhejiang province of China were selected as study areas. The ABCD model and Eckhardt method were used to calculate baseflow and baseflow index (BFI). The temporal and spatial evolution patterns of baseflow were analyzed through statistical analysis and the Mann–Kendall test. The results showed that the ABCD model performs well in simulating overall hydrological processes on the monthly streamflow at BAZ and SD stations with NSE (Nash–Sutcliffe Efficiency) values of 0.82 and 0.83 and Pbias (Percentage Bias) values of 9.2% and 8.61%, respectively. The spatial–temporal distribution of the BFI indicates the higher baseflow contribution in upstream areas compared to downstream areas at both stations. The baseflow and BFI had significant upward trends at the BZA and SD stations in the dry season, while their trends were not uniform during the wet period. These findings are essential guidance for water resource management in the JRB regions.

**Keywords:** monthly-scale hydrological modeling; baseflow simulation; ABCD model; Eckhardt method; spatial and temporal analysis

## 1. Introduction

Baseflow is a relatively stable streamflow component, mainly originating from subsurface streamflow or delayed portions of streamflow [1–3]. Baseflow can provide the necessary support for streamflow replenishment [4] and play a crucial role in supplementing groundwater in the basin [5]. Current global climate change and human activities have increased uncertainty in the hydrological cycle [6,7]. Many studies have analyzed the response of streamflow to climate change and human activities [8,9]. As a relatively stable water source, baseflow is essential in mitigating the impacts of seasonal and climatic variations on water resources [10,11]. A better understanding of groundwater replenishment and discharge processes can be achieved through temporal and spatial baseflow analysis, supporting the scientific and rational utilization of groundwater resources [12]. Therefore, gaining in-depth insights into the temporal and spatial evolution patterns of baseflow in watersheds is crucial for maintaining ecological flow [13,14], managing water resources [15], and studying the patterns of drought evolution [16–18].

Currently, scholars use various baseflow simulation methods [19–21], mainly including numerical simulation methods [22,23], isotope methods [24], and water balance methods [25]. For example, Song et al. [26] quantitatively assessed the contributions of direct streamflow and baseflow to nitrogen loading in the Western Lake Erie Basins in China using numerical simulation methods, which can provide critical information for regional water resource management. Fillo et al. [24] used isotopic techniques to investigate how lawn watering affected baseflow in Denver, USA. They investigated the effects of lawn irrigation on baseflow in semiarid metropolitan regions and concentrated on examining the isotopic ratios of water molecules. Murray et al. [27] conducted an in-depth analysis of monthly baseflow trends in watersheds within Canada using the water balance method. This study aimed to explore the effects of climate change on baseflow, providing critical insights into the changes in hydrological cycles in Canada.

Water balance models can estimate relationships such as precipitation, snowmelt, evaporation, streamflow, and groundwater recharge [28]. The ABCD model, as a four-parameter hydrological model, uses precipitation and potential evapotranspiration as inputs to estimate changes in evapotranspiration, streamflow, soil moisture, and groundwater storage [29]. These four variables play crucial roles in simulating the hydrological processes of a watershed. Model parameter calibration is essential in applying the ABCD model, optimizing model parameters based on observed data to improve predictive accuracy. Standard parameter calibration methods include trial and error by Bayesian methods [30–32], which effectively adjust model parameters to adapt to the characteristics of different watersheds [33–35].

The choice to study the Jiaojiang River Basin is motivated by its significant importance for water resource management and ecological conservation despite facing challenges such as insufficient mainstream monitoring stations and lacking hydrological data for sub-watersheds. In recent years, the basin has experienced water scarcity due to the impacts of climate change and human activities. Therefore, this research aims to apply the monthly water balance four-parameter hydrological model (ABCD model) to (a) calculate streamflow in data-scarce watersheds, (b) simulate baseflow in each sub-watershed, and (c) assess the temporal and spatial evolution patterns of baseflow within the watershed. This study also aims to (d) provide a more scientific basis for future water resource management and ecological conservation as a powerful tool for technical support for regional water resource assessment and management.

## 2. Materials and Methods

### 2.1. Study Area

The JRB is one of the eight essential river basins in the Zhejiang Province, China. It borders the East China Sea to the east and is situated between 120°17′6″ E to 121°41′00″ E longitude and 28°32′2″ N to 29°20′29″ N latitude, covering an area of 6603 km$^2$ [36], as shown in Figure 1. The basin falls within a subtropical monsoon climate zone characterized by four seasons, mild temperatures, and abundant rainfall. The average annual precipitation is 1652 mm, and the average evaporation is 1237.9 mm (observed using a φ20 cm evaporation pan). Precipitation distribution is higher in mountainous areas than plains, with the southern part receiving more rainfall than the northern part. The southwestern and northwestern mountainous regions have the highest precipitation, while the lowest precipitation occurs in the coastal plain area of JRB. In recent years, water scarcity in the Jiaojiang River Basin has been caused by climate change and human activities. Therefore, studying the temporal and spatial evolution patterns of baseflow within the basin can offer technical support for regional water resource assessment and management.

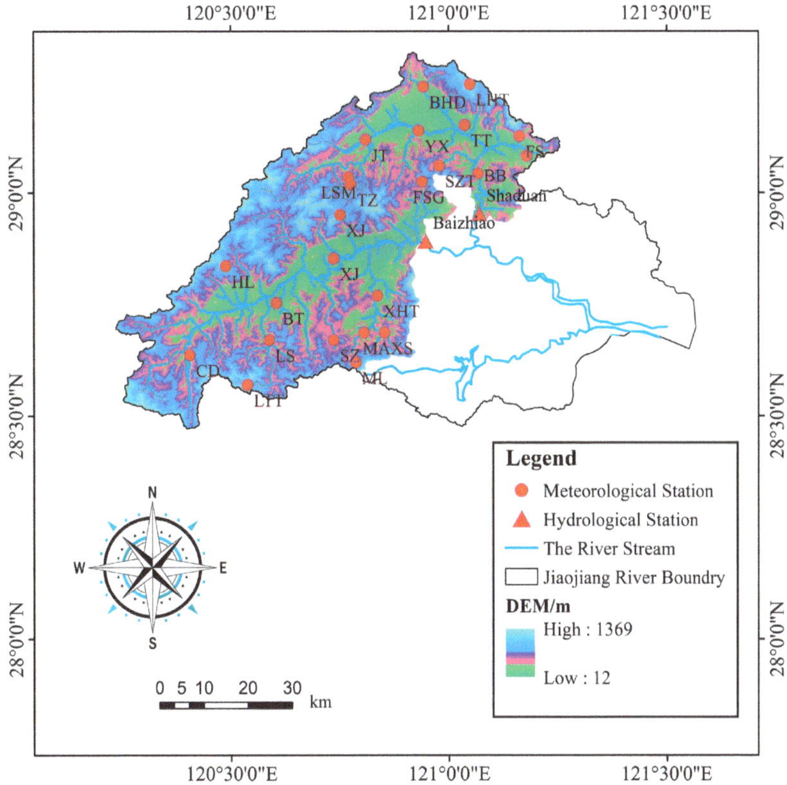


**Figure 1.** Schematic of the study area.

## 2.2. Data

This study selected streamflow data from two hydrological stations, namely, BZA and SD, in the JRB and conducted an analysis. The chosen data period for both stations is unified from 1990 to 2020. The daily precipitation and evaporation pan observation data for 24 meteorological stations within the JRB used in this study were sourced from the China Meteorological Science Data Sharing Service website (http://data.cma.cn, accessed on 1 May 2021).

## 2.3. Methodology

This study employed the Eckhardt [37] method for separating streamflow to obtain baseflow. The simulation of baseflow results is evaluated based on the BFI. Streamflow was simulated based on the ABCD model, and the temporal and spatial baseflow evolution patterns were analyzed through statistical analysis and the Mann–Kendall test [38], as shown in Figure 2.

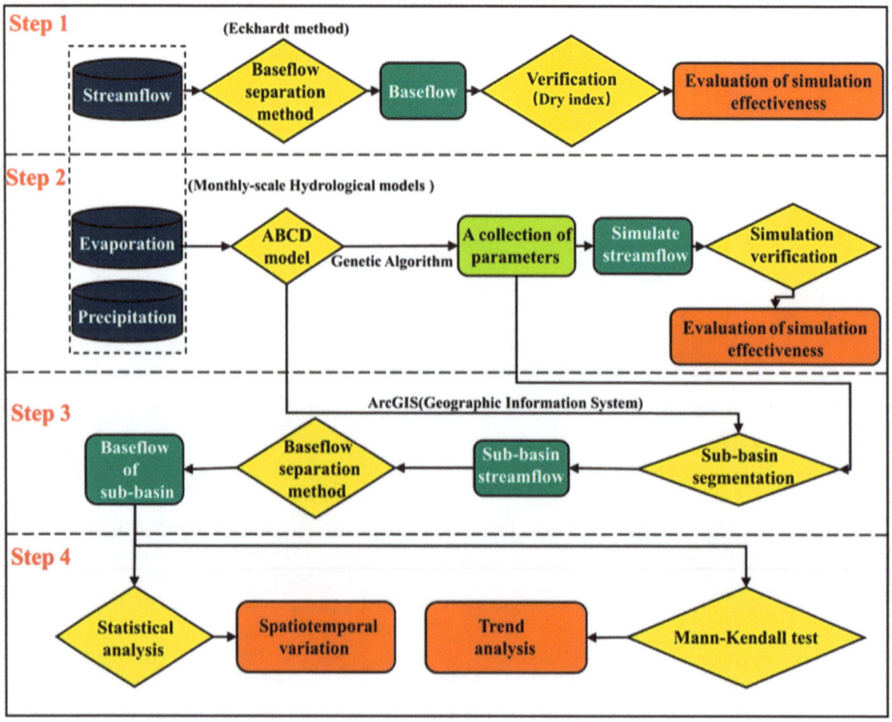


**Figure 2.** Study methodology flow chart.

### 2.3.1. Baseflow Separation Using Digital Filtering Method

The digital filter distinguishes between high-frequency and low-frequency signals because the rapid response qualities of direct streamflow are comparable to high-frequency signals, while the sluggish response characteristics of baseflow are similar to low-frequency signals [39]. Streamflow processes can be divided into direct and baseflow according to this split. Eckhardt [37] developed a general form for some filtering techniques, such as the Chapman–Maxwell method [39] and Lyne–Hollick approach [40], building on earlier research on digital filtering techniques.

$$b_i = Ab_{i-1} + By_i \tag{1}$$

In the equation, $b_i$ represents the baseflow for time $i$, $y_i$ represents the total net flow for time $i$, and $A$ and $B$ are functions of the recession coefficient $\alpha$.

Assuming a linear relationship between the outflow and storage of the aquifer, the coefficients A and B can be expressed in terms of two variables: the recession coefficient ($\alpha$) and the maximum BFI (*BFImax*). A more universally applicable digital filtering equation is obtained through derivation, known as the Eckhardt Digital Filtering Method.

$$b_t = \frac{(1 - BFI_{max})\alpha b_{t-1} + (1-\alpha)BFI_{max}Q_t}{1 - \alpha BFI_{max}} \tag{2}$$

where $\alpha$ is the filtering parameter, $BFI_{max}$ is the maximum of BFI, and $Q_t$ is the measured streamflow at time $t$.

According to the research findings of Eckhardt, the parameter $BFI_{max}$ can take empirical values under different hydrogeological conditions: for unconsolidated porous aquifers with perennial rivers, $BFI_{max}$ is taken as 0.80; for unconsolidated porous aquifers with seasonal rivers, *BFImax* is taken as 0.50; and for weakly permeable aquifers with seasonal rivers, *BFImax* is taken as 0.25. The value of $\alpha$ has a relatively small impact on the calculation results and can generally be set between 0.95 and 0.98.

To implement the Eckhardt method without conducting a hydrogeological investigation, Fan et al. [13] proposed a reverse filtering method using recession constants to calculate *BFImax*:

$$b_{t-1} = \frac{b_t}{\alpha}(b_t \leq Q_t) \tag{3}$$

Perform reverse iteration operations on daily flow based on the equation, and then obtain *BFImax* by dividing the maximum possible total baseflow by the total flow. Fan et al. [13] applied reverse filtering to generate different *BFImax* parameters for 1815 watersheds, which reflects the soil heterogeneity and the spatial variability of hydroclimatic variables.

### 2.3.2. Baseflow Index

The baseflow index (BFI) represents the contribution of baseflow to the total streamflow [41], and the following equation expresses it:

$$\text{BFI} = \frac{\int_{t_1}^{t_2} Q_b(t)dt}{\int_{t_1}^{t_2} Q(t)dt} \tag{4}$$

where $Q$ represents the total streamflow volume in cubic meters per second (m$^3$/s); $Q_b$ is the baseflow volume in cubic meters per second (m$^3$/s); $t$ is the time step; and $t_1$ and $t_2$ represent the starting and ending times, respectively.

### 2.3.3. The Low-Flow Index Method

The Low-Flow Index is an important indicator reflecting the characteristics of groundwater supply to river streamflow. $Q_{90}$ and $Q_{50}$ represent the flow rates occurring with frequencies equal to or greater than 90% and 50%, respectively. These flow rates are determined using the daily flow duration curve [42]. The product of the Low-Flow Index ($Q_{90}/Q_{50}$) and the annual total streamflow were taken as the observed value of the annual baseflow to compare with the above baseflow separation estimation results.

### 2.3.4. ABCD Model

The watershed storage area is conceptualized as layers of soil and groundwater according to the ABCD model [29] (Figure 3). Evapotranspiration losses happen when precipitation reaches the soil layer, and the amount of evapotranspiration varies with soil moisture content in a nonlinear empirical manner. There are two components to streamflow: baseflow is released from groundwater, and direct streamflow is the total surface streamflow, followed by subsurface flow from precipitation and soil water. Groundwater can be replenished by soil water seeping downhill. Many studies [31,32] have used the ABCD model to simulate watershed streamflow in various regions and have obtained good simulation accuracy.

Within a finite time step (monthly or yearly), the mass balance equation for soil water is:

$$S_i - S_{i-1} = P_i - E_i - R_i - D_i \tag{5}$$

where $i$ represents the time step; $S_{i-1}$ and $S_i$ are the soil water storage at the beginning and end of the time step; and $P_i$, $E_i$, $R_i$, and $D_i$ represent the precipitation, evapotranspiration, groundwater recharge, and direct streamflow within the time step, respectively.

In the ABCD model, the actual evapotranspiration ($E$) is a function of precipitation ($P$) and soil water storage ($S$) [29]. It defines two state variables for this purpose, namely, effective water content ($W_i$) and potential evapotranspiration ($Y_i$). They can be expressed by the following equations:

$$W_i = P_i + S_{i-1} \tag{6}$$

$$Y_i = E_i + S_i \tag{7}$$

Assuming a nonlinear functional relationship between $Y_i$ and $W_i$, the following is established:

$$Y_i(W_i) = \frac{W_i + b}{2a} - \left[\left(\frac{W_i + b}{2a}\right)^2 - \frac{W_i b}{a}\right]^{0.5} \tag{8}$$

where both $a$ and $b$ are parameters. The range of $a$ is $0 \le a \le 1$, reflecting the sensitivity of $Y_i$ to changes in $W_i$, and $b$ represents the maximum possible value of $Y_i$:

$$S_i = Y_i exp(E_{0i}/b) \tag{9}$$

where $E_{0i}$ represents the potential evapotranspiration.

Utilizing Equations (3) and (4), Equation (1) can be rewritten as:

$$R_i + D_i = W_i - Y_i \tag{10}$$

For the allocation of $R_i$ and $D_i$, the ABCD model further assumes:

$$\begin{cases} R_i = c(W_i - Y_i) \\ D_i = (1 - c)(W_i - Y_i) \end{cases} \tag{11}$$

where $c$ is the third parameter of the model.

The ABCD model simplifies the groundwater layer as a linear reservoir, which can be expressed as:

$$F_i = dG_i \tag{12}$$

where $d$ is the fourth parameter of the model.

Substituting Equation (9) into Equation (2), we obtain:

$$G_i = \frac{[c(W_i - Y_i) + G_{i-1}]}{(1 + d)} \tag{13}$$

Thus, the total streamflow of the watershed $Q_i = D_i + F_i$.

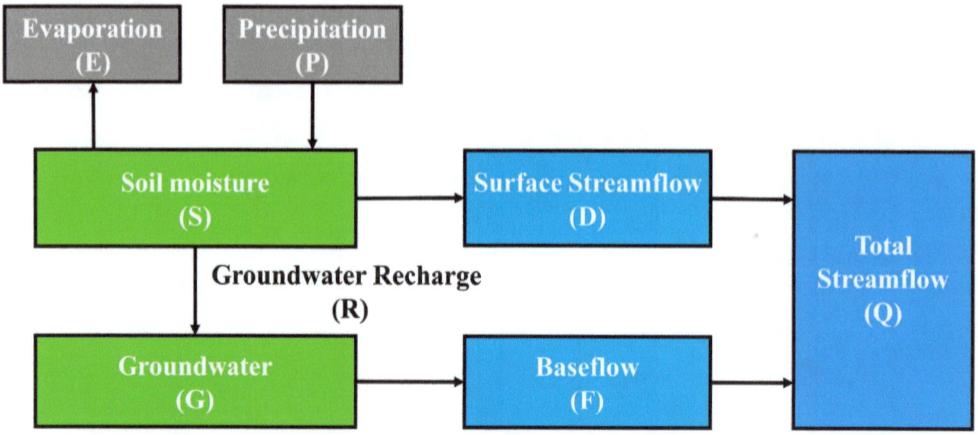


**Figure 3.** Conceptual diagram of the ABCD model.

#### 2.3.5. Model Performance Evaluation

We employed a Genetic Algorithm (GA) for parameter calibration in this study. GA aims to find the optimal solution to a given problem, analogous to Darwin's theory of evolution, where individual traits are preserved in a population. GA preserves a candidate solution set (also known as individuals) tailored to the specific problem. These candidate solutions are iteratively evaluated to create the next generation of solutions. Solutions with better traits are more likely to be selected and pass on their characteristics to the next generation of candidate solutions. As generations progress, the candidate solution set can better address the current problem.

We utilized GA to optimize the parameters of our model to maximize its fit with observed data. Initially, we defined a fitness function to assess the quality of each parameter set. We then initialized a population containing multiple parameter sets as candidate solutions. Subsequently, we iteratively evaluated these candidate solutions and selected the best based on their fitness values to generate the next generation. New solutions were created in each generation through crossover and mutation operations and added to the next generation's population. After multiple iterations, we obtained a set of optimized parameters that maximized the model's fit with the observed data.

The normalized dimensionless NSE (Nash–Sutcliffe Efficiency) [43] compares the variance of the measured and simulated data to determine the fit quality. The NSE value range is −1 to negative infinity. Better simulation results and increased model reliability are indicated by a higher NSE value closer to 1. On the other hand, a lower NSE value, nearer 0, denotes worse simulation outcomes with more significant modeling process mistakes, and the overall dependability of the simulation findings is reduced. The model is deemed untrustworthy if the NSE is noticeably less than 0.

Expressed mathematically as:

$$\text{NSE} = 1 - \frac{\sum_{t=1}^{T}\left(Q_o^t - Q_m^t\right)^2}{\sum_{t=1}^{T}\left(Q_o^t - \overline{Q_o}\right)^2} \tag{14}$$

where $Q_O$ refers to observed values, $Q_m$ refers to simulated values, $Q^t$ denotes a specific value at time $t$, and $\overline{Q_o}$ represents the overall mean of observed values.

The Percentage Bias (Pbias) [44] represents the percentage difference between the measured and simulated water flows compared to the corresponding inferred natural water flow. The closer the Pbias is to 0, the better the model performance.

$$\text{Pbias} = \sum_{i=1}^{n} \frac{Q_{si} - Q_{oi}}{Q_{oi}} \times 100 \tag{15}$$

where $Q_{oi}$ is the observed flow on the $i$-th day; $Q_{si}$ is the simulated flow on the $i$-th day; $Q_o$ is the mean observed flow for the days in question; and $n$ is the length of observed data.

### 2.3.6. Mann–Kendall Test

The World Meteorological Organization suggests the widely used non-parametric Mann–Kendall test [38]. Since Mann and Kendall first put it forth, many studies have used it to examine patterns in time series data of various variables, including temperature, rainfall, streamflow, and water quality. The sample does not need to follow a particular distribution for the Mann–Kendall test to be valid, and it is unaffected by a small number of outliers. Computing appropriately for non-normally distributed data such as meteorological, hydrological, and other types is straightforward.

The null hypothesis H_0 in the Mann–Kendall test states that the time series data $X = (x_1, \ldots, x_n)$ is made up of $n$ independent random variables with identical distributions. The alternative hypothesis $H_1$ is a two-sided test in which the distributions of $x_k$ and $x_j$ are different for every $k, j \leq n$, and $k \neq j$,. The test statistic $S$ is calculated as follows:

$$S = \sum_{k=1}^{n} \sum_{j=k+1}^{n-1} sgn(x_k - x_j) \tag{16}$$

Here, $sgn()$ is the sign function, defined as follows:

$$sgn(x_k - x_j) = \begin{cases} 0 & (x_k - x_j) > 0 \\ 1 & (x_k - x_j) = 0 \\ -1 & (x_k - x_j) < 0 \end{cases} \tag{17}$$

The statistic $S$ follows a normal distribution with a mean of 0, and the variance $Var(S)$ is calculated as follows:

$$Var(S) = \frac{n(n-1)(2n+5)}{18} \tag{18}$$

When $n > 10$, the standard typical statistic is calculated using the following formula:

$$Z = \begin{pmatrix} \frac{S-1}{\sqrt{Var(S)}} & S > 0 \\ 0 & S = 0 \\ \frac{S+1}{\sqrt{Var(S)}} & S < 0 \end{pmatrix} \tag{19}$$

If $Z > 1.96$ or $Z < -1.96$, at a significance level of $\alpha = 0.05$, we reject the null hypothesis $H_0$, indicating enough evidence to support the alternative hypothesis $H1$. If $-1.96 \leq Z \leq 1.96$, at a significance level $\alpha = 0.05$, we accept the null hypothesis $H_0$ [38].

## 3. Results and Discussions

### 3.1. Streamflow Simulation Using ABCD Model

The ABCD model parameters for the BZA and SD hydrological stations are shown in Table 1. The parameter values for the two neighboring basins are relatively close. Once the parameters and initial values were determined, the optimized model parameters were used to optimize the simulated flows from 1990 to 2010. In addition, the initial values of S and G were further optimized to obtain the best simulation results from 2011 to 2020. The correlation with measurements is shown in Figure 4. The simulated results for the BZA and SD hydrological stations have NSE values of 0.82 and 0.83, respectively, and Pbias values of 9.2% and 8.61%. The overall deviation of the water quantities derived from the ABCD model from the observations at the BZA and SD hydrological stations is positive, which suggests that the model slightly overestimates the actual observed values of the mean hydrological processes. The NSE is more than 0.8, and the Pbias is less than 10%, which indicates that the ABCD model is a good fit and can be used for hydrological simulation in JRB [45].

**Table 1.** The ABCD model parameters for two hydrological stations.

| Station | A | B (mm) | C | D | Soil Water Storage (mm) | Groundwater Storage (mm) |
|---|---|---|---|---|---|---|
| BZA | 0.98 | 220.43 | 0.37 | 0.72 | 50.64 | 578.29 |
| SD | 0.96 | 283.69 | 0.30 | 0.60 | 27.50 | 675.55 |

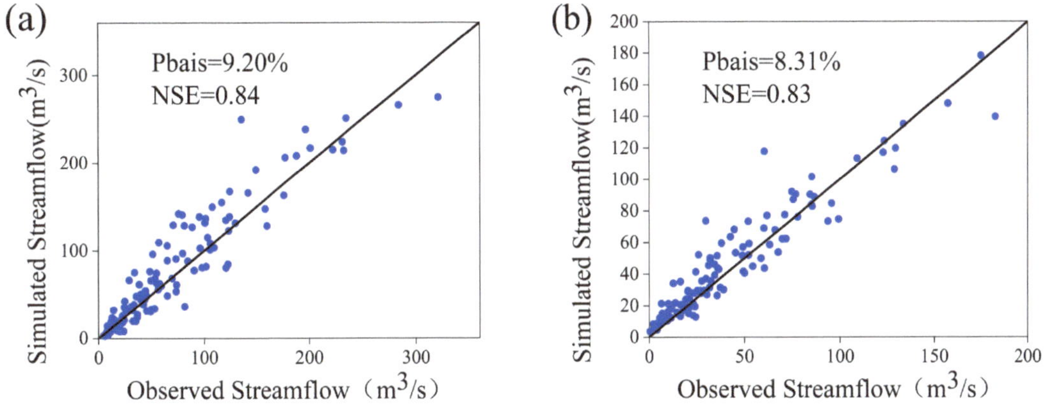


**Figure 4.** Scatter plot of observed and simulated monthly streamflow in the (**a**) BZA and (**b**) SD stations.

### 3.2. Baseflow Simulation Applying Eckhardt Digital Filtering Method

Table 2 illustrates the ABCD model parameters for the BZA and SD hydrological stations. There are slight differences in the parameter values between the two adjacent basins, with the parameter values for BZA consistently higher than those for SD. The parameters at the daily scale exhibit significant fluctuations in both basins. Hence, monthly scale parameters were adopted for excellent stability and reliability in modeling.

**Table 2.** Baseflow and streamflow characteristics at two hydrological stations.

| Station | Average Daily Streamflow ($m^3$/s) | Average Daily Baseflow ($m^3$/s) | Maximum Daily Baseflow ($m^3$/s) | Median Daily Baseflow ($m^3$/s) | Average Monthly Streamflow ($m^3$/s) | Average Monthly Baseflow ($m^3$/s) | Monthly Maximum Baseflow ($m^3$/s) | Median Monthly Baseflow ($m^3$/s) |
|---|---|---|---|---|---|---|---|---|
| BZA | 71.55 | 35.92 | 396.14 | 22.73 | 70.99 | 35.82 | 158.42 | 28.48 |
| SD | 39.33 | 21.81 | 195.80 | 15.31 | 41.36 | 21.72 | 85.72 | 16.82 |

The baseflow simulation was conducted for the SD and BZA watersheds using the ABCD model, followed by baseflow separation using the Eckhardt digital filtering method. To validate the baseflow separation estimation results, we used the product of the Low-Flow Index ($Q_{90}/Q_{50}$) and the annual total streamflow as the observed value of the annual baseflow. This observed value was compared with the baseflow calculated using the ABCD model. The NSE and Pbias were then used to validate the baseflow calculated by the ABCD model. Subsequently, the simulated results were evaluated using the observed streamflow multiplied by the low-flow index (Figure 5). The NSE values for baseflow simulation at the BZA and SD stations were 0.81 and 0.85, respectively. The Pbias values of baseflow simulation for the BZA and SD stations were 8.65% and 5.90%, respectively, indicating a slight underestimation of the observed values by the model. The baseflow separation results showed that this method could effectively simulate baseflow, which can provide a fundamental reflection of the annual variation within the watershed. Moreover, this approach could be utilized to analyze and research the spatiotemporal evolution of baseflow in the SD and BZA stations [46].

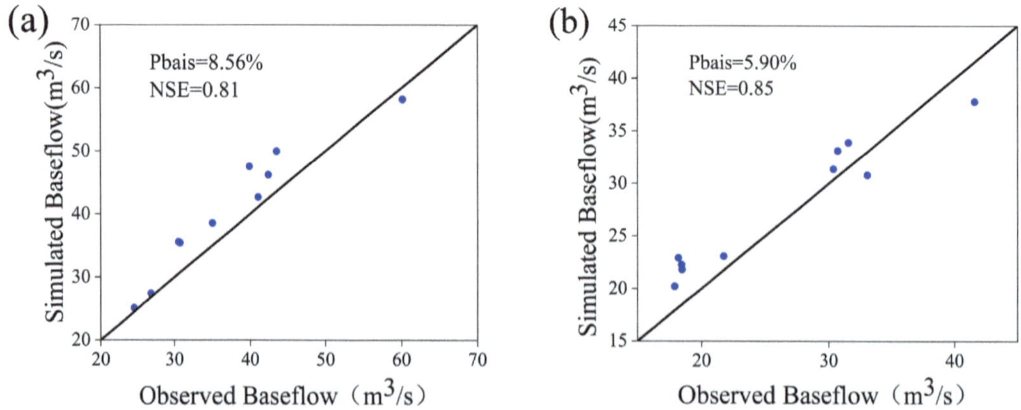


**Figure 5.** Watershed observation and simulation of yearly Baseflow verification in the (**a**) BZA and (**b**) SD stations.

### *3.3. Division of Wet and Dry Periods by BFI Value*

Figure 6 illustrates the variation of monthly BFI based on the results of the digital filtering method to derive BFI values. The BFI variation pattern at the BZA and SD stations remained consistent throughout the year, with the maximum BFI occurring in December and the minimum in August. The flow in rivers decreased as the seasons changed and rainfall decreased. River replenishment has increasingly relied on subsurface runoff, which has increased the supply of subsurface runoff. Consequently, the BFI has increased as the percentage of subsurface runoff to river replenishment has decreased. This has resulted in an increase in groundwater recharge, which has decreased the amount of groundwater contributing to river flow and raised the BFI. During the summer, river recharge is primarily dependent on precipitation and groundwater. The groundwater input to river flow increases proportionately with rainfall, lowering the BFI [47].

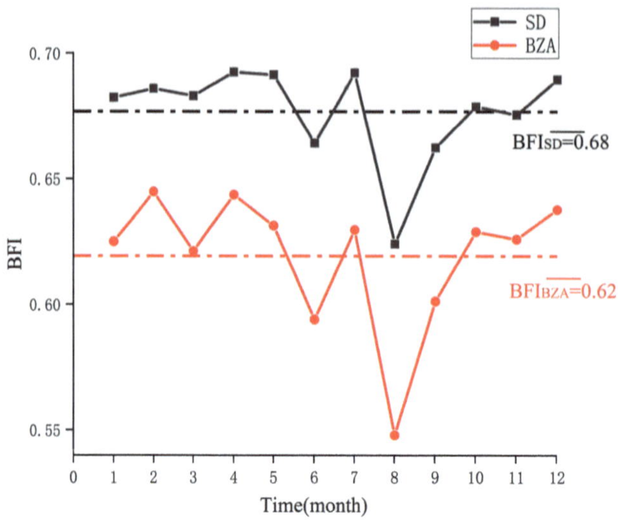


**Figure 6.** Monthly scale BFI changes at BZA and SD stations.

According to the monthly BFI results at the BZA and SD stations, the variation of monthly BFI is shown in Figure 7. By comparing multi-year average BFI values, the period from October to April was identified as the dry season, while May to September was considered the wet season for the watershed. As a result, the rainy season was defined as the months with BFI below average and the dry season as those with BFI above average [48].

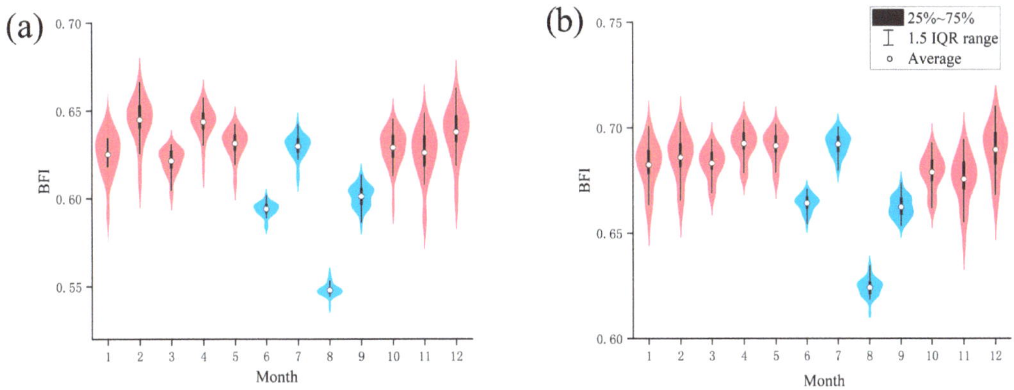


**Figure 7.** A comparison of BFI between wet and dry seasons for (**a**) BAZ and (**b**) SD stations. The red (blue) color represents dry (wet) season.

### 3.4. Spatial and Temporal Distribution Characteristics of Baseflow

The monthly baseflow and BFI statistical analysis for the BZA and SD watersheds from 1990 to 2020 are shown in Figure 8. As indicated in Figure 8, the BFI did not exhibit significant differences due to variations in watershed area, suggesting that the model performed well in capturing the baseflow process. The watershed area considerably impacted baseflow discharge, increasing baseflow as the area expanded. Numerous outliers in the baseflow for both BZA and SD suggested significant fluctuations in monthly flow rates [47].

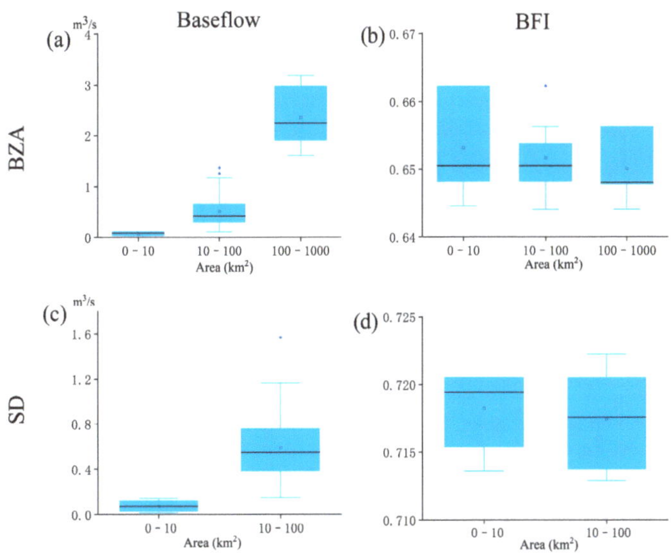


**Figure 8.** The boxplots of baseflow and BFI for (**a**,**b**) BAZ and (**c**,**d**) SD stations.

The baseflow in the JRB exhibited a significant geographical gradient trend, increasing from southwest to northeast. There was a notable gradient difference in baseflow among different regions within the basin, ranging from below 0.27 $m^3/s$ to over 3.27 $m^3/s$. This may be attributed to the typically higher elevation or complex terrain in the southwest, which could have led to a more concentrated water flow, forming high-value areas of baseflow. Conversely, the lower terrain in the northeast may have resulted in relatively lower baseflow. The terrain gradient could have accelerated the flow of water, influencing the distribution of baseflow across different basin regions. The seasonal variations and distribution of precipitation in the subtropical monsoon climate could have induced spatial variations in baseflow. Monsoonal climates are often characterized by distinct wet and dry seasons, which may have contributed to differences in baseflow within the basin [44]. Higher terrain in the southwest might have been more susceptible to the influence of monsoon rains, while the northeast could have been relatively drier. WNW winds might have created differences in moisture transport within the basin, resulting in spatial variations in baseflow [46]. Specific wind directions might have transported moisture to specific areas of the basin, affecting precipitation and thus influencing baseflow formation. These factors collectively may have contributed to the spatial variation of baseflow within the basin.

Figure 9 showed the BFI, baseflow modulus, and average annual baseflow distribution for all Sub-watersheds at BZA and SD basins during all, wet, and dry seasons. The BZA watershed has a spatial trend in the baseflow modulus, with larger values in the upstream areas and smaller values in the downstream areas. The baseflow modulus in the upstream regions exhibits a broader range of fluctuations, ranging from 140 to over 220 (L/km·s), while it is relatively minor in the downstream areas. Geological conditions and hydrological processes in the upstream areas may have influenced the spatial differences in the baseflow modulus. However, the baseflow modulus was significant in upstream and downstream areas, with a relatively large fluctuation range maintained between 100 and 180 (L/km·s) in the SD watershed. The study suggests complex hydrological processes governing baseflow in the entire SD watershed. Factors such as relatively flat terrain similarly affect the basin's baseflow formation. With comparatively more significant values in the upstream areas ranging from below 0.66 to over 0.68, the BFI in the BZA watershed is generally smaller. By contrast, the BFI in the SD watershed is often more significant, fluctuating within an overall range of 0.68 to 0.74. Precipitation, evaporation, terrain, and other complicated elements within the basin may have impacted the spatial differences in the BFI among different locations [49–51].

Figure 10 illustrates the overall baseflow trend in the JRB. Both the BZA and SD areas exhibit significant increases in baseflow, with a slight increase in the baseflow index upstream of BZA and a slight decrease downstream. During the wet season, the baseflow in BZA generally decreases significantly, while the baseflow index in BZA increases significantly, and the SD area shows an overall significant decrease. In contrast, during the dry season, the baseflow and baseflow index in all areas exhibit significant increases. The trends of average baseflow discharge and BFI for all sub-basins during the wet and dry seasons indicate an annual increase in baseflow, suggesting the basin is building up groundwater reserves. Groundwater responds slowly to changes due to its lengthy recharging cycle. Therefore, the increase in baseflow can indicate substantial groundwater recharge in earlier periods, contributing to steady growth in groundwater reserves [52]. Furthermore, the yearly increase in the BFI has demonstrated an upward tendency in the baseflow proportion relative to the overall streamflow. Due to the subtropical monsoon climate and topography of the JRB, there may be a more significant percentage of groundwater recharging with total streamflow, as shown by this rise in the BFI. There were notable seasonal and regional variations in temperature and precipitation distribution.

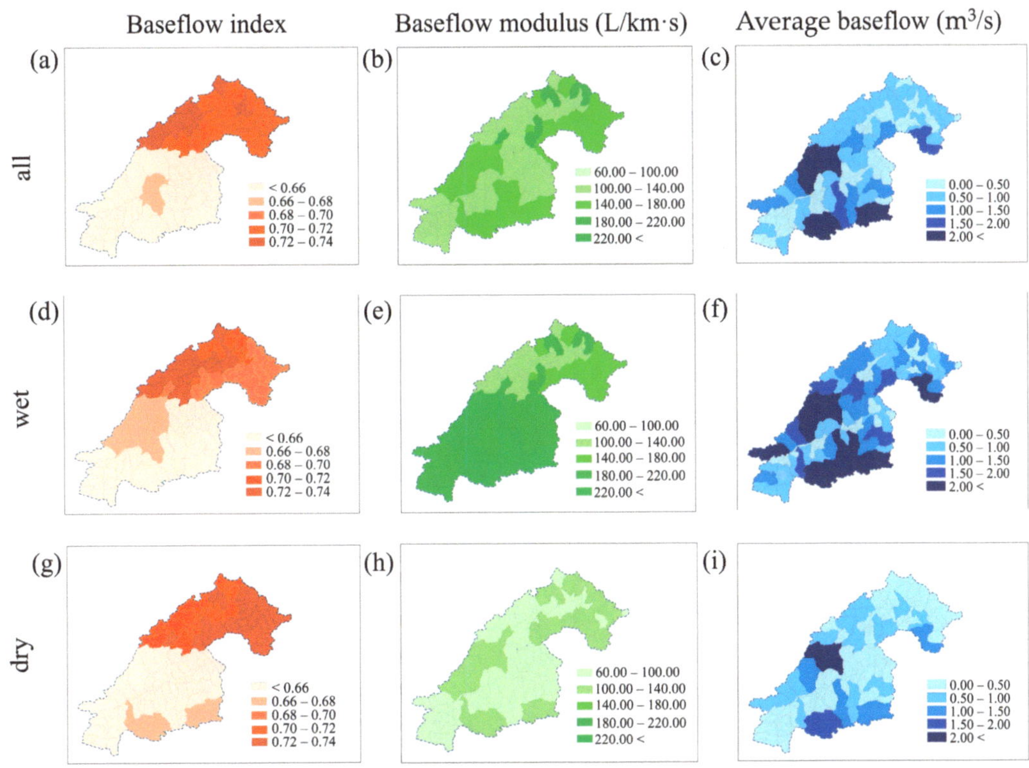


**Figure 9.** BFI, baseflow modulus, and average annual baseflow distribution for all Sub-Watersheds in (**a–c**) all seasons, (**d–f**) wet season, and (**g–i**) dry season.

Figure 11 illustrates the correlation analysis between evaporation, precipitation, and baseflow for all sub-basins. Surface runoff and groundwater infiltration may be impacted by changes in precipitation patterns and temperature distribution brought on by climate change. Long-term climate patterns may have increased groundwater recharge and baseflow generation. Groundwater recharge and circulation may have been significantly impacted by the basin's geological features [53]. Precipitation and baseflow are generally positively correlated, while the BFI is negatively correlated. The differences between the wet and dry seasons are not significant overall. Evaporation is generally positively correlated with the baseflow and BFI. During the wet season, evaporation negatively correlates with the baseflow and strongly positively correlates with the BFI. During the dry season, evaporation positively correlates with the baseflow and weakly positively correlates with the BFI. Overall, the baseflow in the BZA is more positively correlated with evaporation than in the SD. The SD area is generally more negatively correlated with the BFI than the BZA area. There is not much difference in the correlation of evaporation with the baseflow and the BFI between the two basins.

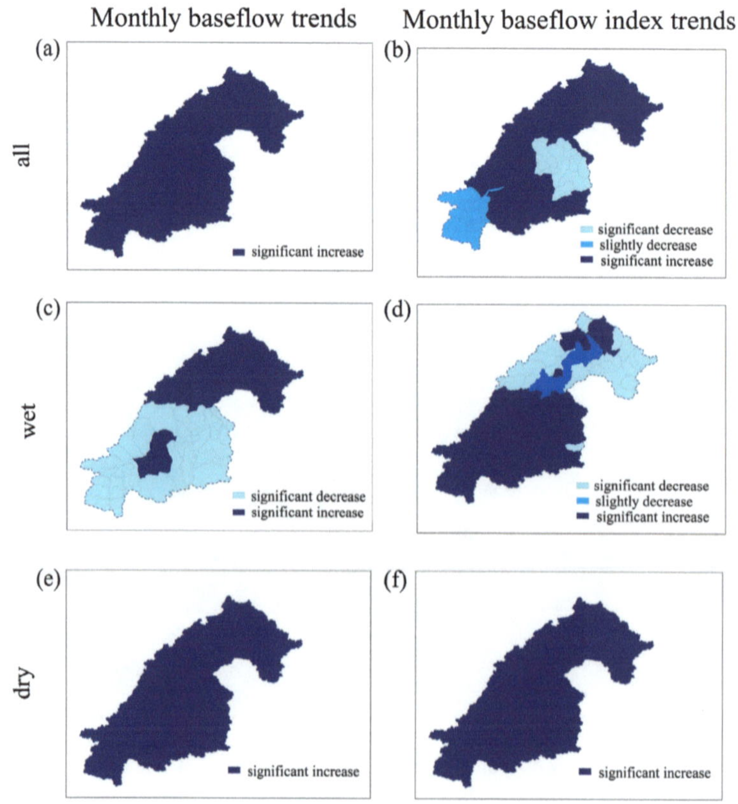


**Figure 10.** The trend distribution of average baseflow discharge, BFI, and baseflow modulus for all sub-basins in (**a**,**b**) all seasons, (**c**,**d**) wet season, and (**e**,**f**) dry season.

The genesis of baseflow may have been impacted by different groundwater flow patterns in the aquifers due to variations in the geological structure and rock types found in the JRB [51]. During the rainy season, baseflow increased somewhat, while the patterns varied by region. During this time, the BFI somewhat dropped, suggesting a decline in the proportion of groundwater in the overall streamflow. During the rainy season, there was a noticeable increase in streamflow discharge, which may have approached the maximum baseflow discharge. This implied that while the percentage of baseflow reduced during heavy rainfall, surface streamflow contributed comparatively more to the overall streamflow. Throughout the dry season, baseflow showed a significant increasing trend, whereas the rise of BFI remained pretty stable. One possible explanation for the sharp rise in baseflow during the dry season is a combination of reduced precipitation, comparatively low evapotranspiration, and steady groundwater recharging. The percentage of baseflow in the overall streamflow grew during this period, indicating the growing importance of groundwater in preserving the basin's hydrological balance [54].

The study is constrained by limited runoff data, which may introduce uncertainties in the simulation accuracy and the absence of measured baseflow data, leading to uncertainties in the baseflow simulation. Future research could include field experiments for validating baseflow simulations and conducting attribution analysis to identify critical factors, such as climate change and human activities, influencing baseflow changes in the watershed.

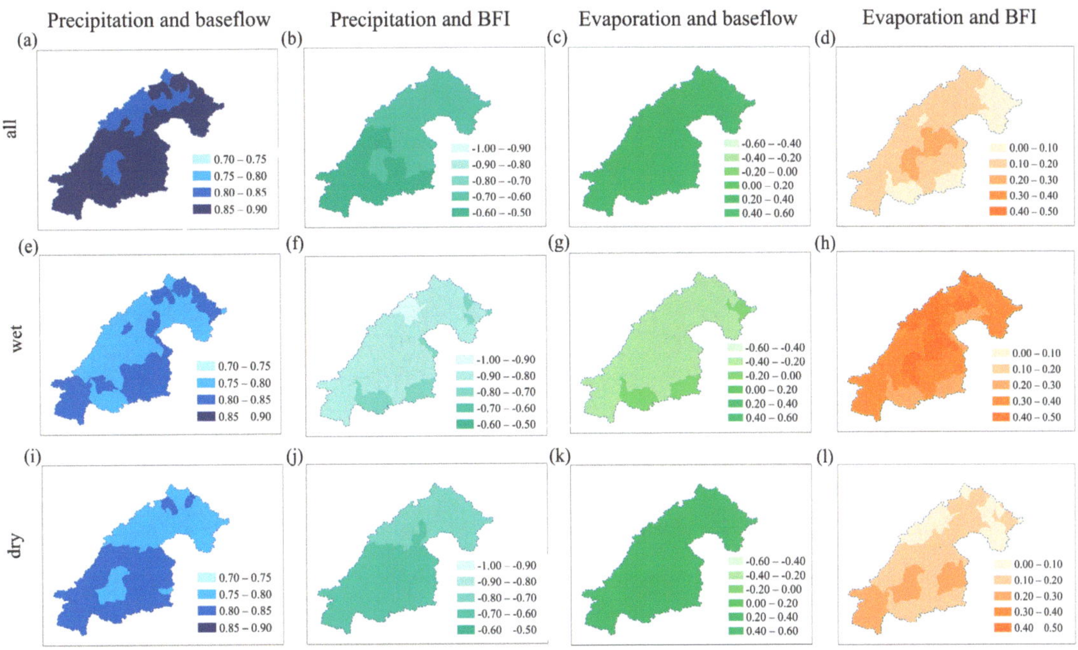


**Figure 11.** Correlation analysis between evaporation, precipitation, and baseflow for all sub-basins in (**a–d**) all seasons, (**e–h**) wet season, and (**i–l**) dry season.

## 4. Conclusions

This study utilized monthly streamflow data from the BZA and SD stations in the JRB to simulate baseflow (BFI) using the Eckhardt method and the ABCD model. In addition, it has achieved baseflow simulation in data-scarce basins based on a monthly scale hydrological model. The primary findings from this study are as follows:

(1) The simulated findings show that the NSE values for the BZA and SD stations are 0.82 and 0.83, and the Pbias values are 9.2% and 8.61%. According to available data, the ABCD model generally replicates monthly hydrological processes but overestimates the streamflow at the BZA and SD basins.

(2) Using the Eckhardt method to separate baseflow, the NSE values of baseflow simulations at the BZA and SD stations were 0.81 and 0.85, respectively. The Pbias values were 8.65% and 5.90%, respectively, which indicates that the model slightly overestimates baseflow in the BZA and SD stations. According to the BFI spatial distribution, there is a trend toward greater values in the upstream regions and lower values in the downstream regions. The BFI rises yearly, and the monthly hydrological model's baseflow exhibits a relatively fast expansion pattern.

(3) The increasing trends of baseflow were relatively small during the wet season but more significant during the dry season, highlighting the impact of seasonal variations on baseflow simulation in the monthly-scale hydrological model. The baseflow modulus in the upstream regions shows a broader range of fluctuations from 140–220 (L/km·s) and 100–180 (L/km·s) at the BZA and SD stations, respectively. Geological conditions and hydrological processes in the upstream areas may have influenced the spatial differences in the baseflow modulus.

This study provided a detailed analysis and validation for baseflow simulation in data-scarce basins using monthly-scale hydrological models. These findings are essential for gaining deeper insights into applying monthly-scale hydrological models in data-scarce basins, scientifically managing water resources, providing environmental protection, and adapting to climate change.

**Author Contributions:** Conceptualization, D.X.; methodology, H.H. and Y.C.; software, H.H.; validation, H.H. and F.Z.; formal analysis, H.H., H.X. and D.X.; investigation, H.H., D.X. and Z.Z.; resources, Y.C. and B.X.; data curation, D.X.; writing—original draft preparation, H.H.; writing—review and editing, H.H., F.Z. and D.X.; visualization, H.H.; supervision, H.H.; project administration, H.N.; and funding acquisition, H.X. All authors have read and agreed to the published version of the manuscript.

**Funding:** This research was funded by the Zhejiang Provincial Water Resources Department Science and Technology Program, grant numbers RC2201 and RB2002; the Natural Science Foundation of Zhejiang Province, grant number LZJWY22E090007; and the Scientific Research Fund of the Zhejiang Provincial Education Department, grant number Y202352492.

**Data Availability Statement:** The daily precipitation and evaporation pan observation data for 24 meteorological stations within the JRB used in this study are sourced from the China Meteorological Science Data Sharing Service website (http://data.cma.cn, accessed on 1 May 2021).

**Acknowledgments:** We are grateful to the Zhejiang Hydrological Management Center for providing hydrological data.

**Conflicts of Interest:** Author Donghui Xie was employed by the company Ningbo Yuanshui Company Limited. The remaining authors declare that the research was conducted in the absence of any commercial or financial relationships that could be construed as a potential conflict of interest.

## References

1. Chen, H.; Huang, S.; Xu, Y.P.; Teegavarapu, R.S.V.; Guo, Y.; Nie, H.; Xie, H. Using baseflow ensembles for hydrologic hysteresis characterization in humid basins of Southeastern China. *Water Resour. Res.* **2024**, *60*, e2023WR036195. [CrossRef]
2. Tallaksen, L.M. A review of baseflow recession analysis. *J. Hydrol.* **1995**, *165*, 349–370. [CrossRef]
3. Singh, S.K.; Pahlow, M.; Booker, D.J.; Shankar, U.; Chamorro, A. Towards BFI characterisation at national scale in New Zealand. *J. Hydrol.* **2019**, *568*, 646–657. [CrossRef]
4. Miller, M.P.; Buto, S.G.; Susong, D.D.; Rumsey, C.A. The importance of Baseflow in sustaining surface water flow in the Upper Colorado River Basin. *Water Resour. Res.* **2016**, *52*, 3547–3562. [CrossRef]
5. Niazi, A.; Bentley, L.R.; Hayashi, M. Estimation of spatial distribution of groundwater recharge from stream baseflow and groundwater chloride. *J. Hydrol.* **2017**, *546*, 380–392. [CrossRef]
6. Huntington, T.G. Evidence for intensification of the global water cycle: Review and synthesis. *J. Hydrol.* **2006**, *319*, 83–95. [CrossRef]
7. Pendergrass, A.G.; Knutti, R.; Lehner, F.; Deser, C.; Sanderson, B.M. Precipitation variability increases in a warmer climate. *Sci. Rep.* **2017**, *7*, 17966. [CrossRef] [PubMed]
8. Milly, P.C.; Dunne, K.A. Colorado River flow dwindles as warming-driven loss of reflective snow energizes evaporation. *Science* **2020**, *367*, 1252–1255. [CrossRef] [PubMed]
9. Vano, J.A.; Udall, B.; Cayan, D.R.; Overpeck, J.T.; Brekke, L.D.; Das, T.; Lettenmaier, D.P. Understanding uncertainties in future Colorado River streamflow. *Bull. Am. Meteorol. Soc.* **2014**, *95*, 59–78. [CrossRef]
10. Eltahir, E.A.B.; Yeh, P. On the asymmetric response of aquifer water level to floods and droughts in Illinois. *Water Resour. Res.* **1999**, *35*, 1199–1217. [CrossRef]
11. Peters, E.; Torfs, P.; Van Lanen, H.A.J.; Bier, G. Propagation of drought through groundwater—A new approach using linear reservoir theory. *Hydrol. Process.* **2003**, *17*, 3023–3040. [CrossRef]
12. Ciampittiello, M.; Dresti, C.; Saidi, H. Water resource management through understanding of the water balance components: A case study of a Sub-Alpine Shallow Lake. *Water* **2021**, *13*, 3124. [CrossRef]
13. Fan, Y.; Li, H.; Miguez-Macho, G. Global patterns of groundwater table depth. *Science* **2013**, *339*, 940–943. [CrossRef] [PubMed]
14. Poff, L.R.; Allan, J.D.; Bain, M.B.; Karr, J.R.; Prestegaard, K.L.; Richter, B.D.; Sparks, R.E.; Stromberg, J.C. The natural flow regime. *BioScience* **1997**, *47*, 769–784. [CrossRef]
15. Ebrahim, G.Y.; Villholth, K.G. Estimating shallow groundwater availability in small catchments using streamflow recession and instream flow requirements of rivers in South Africa. *J. Hydrol.* **2016**, *541*, 754–765. [CrossRef]
16. Sophocleous, M. Interactions between groundwater and surface water: The state of the science. *Hydrogeol. J.* **2002**, *10*, 52–67. [CrossRef]
17. Ahiablame, L.; Sheshukov, A.Y.; Rahmani, V.; Moriasi, D. Annual baseflow variations as influenced by climate variability and agricultural land use change in the Missouri River Basin. *J. Hydrol.* **2017**, *551*, 188–202. [CrossRef]
18. Beck, H.E.; Van Dijk, A.I.; Miralles, D.G.; De Jeu, R.A.; Bruijnzeel, L.A.; McVicar, T.R.; Schellekens, J. Global patterns in BFI and recession based on streamflow observations from 3394 catchments. *Water Resour. Res.* **2013**, *49*, 7843–7863. [CrossRef]

19. George, P.M.; Sekhar, M. Base flow simulation using a physically based subsurface model–The case of a tropical basin in the Western Ghats, India. *J. Hydrol.* **2022**, *613*, 128451. [CrossRef]
20. Luo, Z.; Shao, Q. A modified hydrologic model for examining the capability of global gridded PET products in improving hydrological simulation accuracy of surface runoff, streamflow and baseflow. *J. Hydrol.* **2022**, *610*, 127960. [CrossRef]
21. Longobardi, A.; Villani, P. Baseflow index characterization in typical temperate to dry climates: Conceptual analysis and simulation experiment to assess the relative role of climate forcing features and catchment geological settings. *Hydrol. Res.* **2023**, *54*, 136–148. [CrossRef]
22. Rumsey, C.A.; Miller, M.P.; Sexstone, G.A. Relating hydroclimatic change to streamflow, baseflow, and hydrologic partitioning in the Upper Rio Grande Basin, 1980 to 2015. *J. Hydrol.* **2020**, *584*, 124715. [CrossRef]
23. Welty, C.; Moore, J.; Bain, D.J.; Talebpour, M.; Kemper, J.T.; Groffman, P.M.; Duncan, J.M. Spatial heterogeneity and temporal stability of baseflow stream chemistry in an urban watershed. *Water Resour. Res.* **2023**, *59*, e2021WR031804. [CrossRef]
24. Fillo, N.K.; Bhaskar, A.S.; Jefferson, A.J. Lawn irrigation contributions to semiarid urban baseflow based on water-stable isotopes. *Water Resour. Res.* **2021**, *57*, e2020WR028777. [CrossRef]
25. Delesantro, J.M.; Duncan, J.M.; Riveros-Iregui, D.; Blaszczak, J.R.; Bernhardt, E.S.; Urban, D.L.; Band, L.E. The Nonpoint Sources and Transport of Baseflow Nitrogen Loading Across a Developed Rural-Urban Gradient. *Water Resour. Res.* **2022**, *58*, e2021WR031533. [CrossRef]
26. Song, J.H.; Her, Y.; Guo, T. Quantifying the contribution of direct streamflow and baseflow to nitrogen loading in the Western Lake Erie Basins. *Sci. Rep.* **2022**, *12*, 9216. [CrossRef] [PubMed]
27. Murray, J.; Ayers, J.; Brookfield, A. The impact of climate change on monthly baseflow trends across Canada. *J. Hydrol.* **2023**, *618*, 129254. [CrossRef]
28. Milly, P.C.; Betancourt, J.; Falkenmark, M.; Hirsch, R.M.; Kundzewicz, Z.W.; Lettenmaier, D.P.; Stouffer, R.J. Stationarity is dead: Whither water management? *Science* **2008**, *319*, 573–574. [CrossRef]
29. Thomas, H. *Improved Methods for National Water Assessment: Final Report*; U.S. Geological Survey: Reston, VA, USA, 1981; Volume 44. [CrossRef]
30. Wang, X.; Gao, B.; Wang, X. A modified ABCD model with temperature-dependent parameters for cold regions: Application to reconstruct the changing streamflow in the headwater catchment of the Golmud River, China. *Water* **2020**, *12*, 1812. [CrossRef]
31. Zou, Y.; Yan, B.; Feng, B.; Zhang, J.; Tang, Y.A. Three-Parameter Hydrological Model for Monthly Runoff Simulation—A Case Study of Upper Hanjiang River Basin. *Water* **2023**, *15*, 474. [CrossRef]
32. Leng, J.; Ma, K.; Gu, S.; Zhang, K.; He, D. A non-stationary impactquant framework for assessing the human activity impacts on hydrological drought in the Upper Red River. *Atmos. Res.* **2024**, *304*, 107419. [CrossRef]
33. Ghiggi, G.; Humphrey, V.; Seneviratne, S.I.; Gudmundsson, L. G-RUN ENSEMBLE: A multi-forcing observation-based global streamflow reanalysis. *Water Resour. Res.* **2021**, *57*, e2020WR028787. [CrossRef]
34. Do, H.X.; Gudmundsson, L.; Leonard, M.; Westra, S. The Global Streamflow Indices and Metadata Archive (GSIM)–Part 1: The production of a daily streamflow archive and metadata. *Earth Syst. Sci. Data* **2018**, *10*, 765–785. [CrossRef]
35. Hrachowitz, M.; Savenije, H.H.G.; Blöschl, G.; McDonnell, J.J.; Sivapalan, M.; Pomeroy, J.W.; Arheimer, B.; Blume, T.; Clark, M.P.; Ehret, U.; et al. A decade of Predictions in Ungauged Basins (PUB)—A review. *Hydrol. Sci. J.* **2013**, *58*, 1198–1255. [CrossRef]
36. Chen, H.; Huang, S.; Xu, Y.-P.; Teegavarapu, R.S.V.; Guo, Y.; Nie, H.; Xie, H.; Zhang, L. River ecological flow early warming forecasting using baseflow separation and machine learning in the Jiaojiang River Basin, Southeast China. *Sci. Total Environ.* **2023**, *882*, 163571. [CrossRef] [PubMed]
37. Eckhardt, K. How to construct recursive digital filters for baseflow separation. *Hydrol. Process.* **2005**, *19*, 507–515. [CrossRef]
38. Hamed, K.H.; Rao, A.R. A modified Mann-Kendall trend test for autocorrelated data. *J. Hydrol.* **1998**, *204*, 182–196. [CrossRef]
39. Chapman, T.G. Comment on "Evaluation of automated techniques for Baseflow and recession analyses" by RJ Nathan and TA McMahon. *Water Resour. Res.* **1991**, *27*, 1783–1784. [CrossRef]
40. Lyne, V.; Hollick, M. *Stochastic Time-Variable Rainfall-Streamflow Modelling*; Institute of Engineers Australia: Barton, Australia, 1979; Volume 79, pp. 89–93. Available online: https://www.researchgate.net/publication/272491803_Stochastic_TimeVariable_Rainfall-Streamflow_Modeling (accessed on 1 June 2022).
41. Chen, H.; Teegavarapu, R.S.V.; Xu, Y.-P. Oceanic-Atmospheric Variability Influences on Baseflows in the Continental United States. *Water Resour. Manag.* **2021**, *35*, 3005–3022. [CrossRef]
42. Ayers, J.R.; Villarini, G.; Jones, C.; Schilling, K. Changes in monthly baseflow across the US Midwest. *Hydrol. Process.* **2019**, *33*, 748–758. [CrossRef]
43. Nash, J.E.; Sutcliffe, J.V. River flow forecasting through conceptual models part I—A discussion of principles. *J. Hydrol.* **1970**, *10*, 282–290. [CrossRef]
44. Chen, H.; Xu, Y.P.; Teegavarapu, R.S.; Guo, Y.; Xie, J. Assessing different roles of baseflow and surface streamflow for long-term streamflow forecasting in southeastern China. *Hydrolog. Sci. J.* **2021**, *66*, 2312–2329. [CrossRef]
45. Baseri, M.; Mahjoobi, E.; Rafiei, F.; Baseri, M. Evaluation of ABCD water balance conceptual model using remote sensing data in ungauged watersheds (a case study: Zarandeh, Iran). *Environ. Earth Sci.* **2023**, *82*, 126. [CrossRef]
46. Martinez, G.F.; Gupta, H.V. Toward improved identification of hydrological models: A diagnostic evaluation of the "abcd" monthly water balance model for the conterminous United States. *Water Resour. Res.* **2010**, *46*, W08507. [CrossRef]

47. Bloomfield, J.P.; Gong, M.; Marchant, B.P.; Coxon, G.; Addor, N. How is Baseflow Index (BFI) impacted by water resource management practices? *Hydrol. Earth Syst. Sci.* **2021**, *25*, 5355–5379. [CrossRef]
48. Jiang, H.; Wang, Z.; Ye, A.; Liu, K.; Wang, X.; Wang, L. Hydrological characteristic-based methodology for dividing flood seasons: An empirical analysis from China. *Environ. Earth Sci.* **2019**, *78*, 399. [CrossRef]
49. Chen, S.; Ruan, X. A hybrid Budyko-type regression framework for estimating baseflow from climate and catchment attributes. *J. Hydrol.* **2023**, *618*, 129118. [CrossRef]
50. Saedi, J.; Sharifi, M.R.; Saremi, A.; Babazadeh, H. Assessing the impact of climate change and human activity on streamflow in a semiarid basin using precipitation and baseflow analysis. *Sci. Rep.* **2022**, *12*, 9228. [CrossRef] [PubMed]
51. Aboelnour, M.A.; Engel, B.A.; Frisbee, M.D.; Gitau, M.W.; Flanagan, D.C. Impacts of watershed physical properties and land use on baseflow at regional scales. *J. Hydrol. Reg. Stud.* **2021**, *35*, 100810. [CrossRef]
52. Hellwig, J.; Stoelzle, M.; Stahl, K. Groundwater and baseflow drought responses to synthetic recharge stress tests. *Hydrol. Earth Syst. Sci.* **2021**, *25*, 1053–1068. [CrossRef]
53. Tan, X.; Liu, B.; Tan, X. Global changes in baseflow under the impacts of changing climate and vegetation. *Water Resour. Res.* **2020**, *56*, e2020WR027349. [CrossRef]
54. Mo, C.; Ruan, Y.; Xiao, X.; Lan, H.; Jin, J. Impact of climate change and human activities on the baseflow in a typical karst basin, Southwest China. *Ecol. Indic.* **2021**, *126*, 107628. [CrossRef]

## 1. Introduction

Hydrological processes are affected by several elements such as the land–use type, surface conditions, and climatic and meteorological conditions, which vary spatially and temporally [1–5]. The increasing complexity of hydrological models, coupled with the diversity of data sources, amplifies uncertainties in simulation and prediction. As a result, the prediction of water availability and integrated watershed management becomes a necessary and challenging issue restricted by the implementation of water shortages [6–8]. Meanwhile, the enormous complexities associated with human–environmental interactions make it even more challenging to develop reliable models and schemes to support effective hydrological modelling and water resource management. This dual challenge necessitates innovative solutions and a deeper understanding of the uncertainties involved.

Several scholars have previously applied stochastic analysis and fuzzy mathematics to delve into the intricate uncertainties within water resources systems. Generally, there are three aspects of uncertainties in hydrological modeling: the systematic bias of model input, uncertainty parameters, and structural uncertainty in hydrologic models [9–12]. Notably, parameter uncertainty has been the subject of extensive study, with the Generalized Likelihood Uncertainty Estimation (GLUE) and Bayesian methods serving as commonly utilized tools for evaluating model parameters [13–15]. Both the GLUE and Bayesian methods estimate parameter uncertainty based on likelihood functions [16–19]. As there exist lots of uncertainties for water resources management, the decision makers are usually confronted with challenges to satisfy numerous or contradictory requests [20,21]. The stochastic and fuzzy mathematical programming methods have been adopted by various researchers to address such uncertainties [22–26]. Due to the uncertainties and complexities of research on hydrologic simulation and water resources management, it is essential to keep up with the scientific structure and frontier in a certain domain of science. Then, researchers and decision makers can stay abreast of the latest developments and insights, ensuring the continued advancement of knowledge in hydrological sciences. Recently, an integrated modeling system from hydrological modeling of the natural system to optimization management of the social system has been applied to real–world problems. Zhuang et al. [27] proposed an innovative method that combines simulation and optimization techniques to evaluate the influence of climate change on water resource dynamics. Their findings underscore the substantial impact of uncertainties within the system, which have a pronounced effect on the allocation of water resources, including target and shortage. Li et al. [28] developed an integrated simulation–optimization modeling system for water resources management under the coupled impacts of climate and land use variabilities with priority in ecological protection. The system has tremendous significance for evaluating hydrologic variations with complicated uncertainties and providing optimal water allocation schemes responding to the coupled impacts of climate and land–use variations among society, the economy, and the environment.

Although methodologies developed in previous studies can be effective in addressing various uncertainties in hydrological modeling and water resources management, very little analysis has been carried out from a scientometric and bibliometric perspective. Furthermore, no previous review has provided the development process and the structural relationship of scientific knowledge through visual maps in this field. Therefore, we will use CiteSpace, a graphical tool on account of collaboration, co–citation, and co–occurrence networks, to provide valuable, timely, repeatable, and flexible perspectives to explore the new emerging trends and recognize critical evidence [29–32]. This systematic review utilizes visualization software as a key instrument for mapping the development process and structural relationships of scientific knowledge. In doing so, our research contributes to track the trajectory of emerging trends and pinpoint critical evidence for future advancements in these critical domains.

The main objectives for this research are to supply cooperation, co–citation, and co–occurrence networks with related references obtained from the Web of Science (WOS) Core Collection. Firstly, the study identifies innovative scholars, providing a comprehensive perspective on their contributions from a global and institutional standpoint. Secondly, a distributed network is constructed, delving into the intricate relationships among articles, authors, and journals. Thirdly, through a co–occurrence analysis of keywords and classes, the research unveils pivotal themes and subjects, contributing to the delineation of critical knowledge domains within the field. By employing visual research methodologies, this study goes beyond conventional analyses, scrutinizing the intellectual structure, knowledge characteristics, and research frontiers. In summary, the study contributes to the development of more informed and effective strategies for water system management, especially in addressing challenges such as water scarcity, land use, and climate change. This research not only advances our understanding of hydrological processes but also provides sustainable adaptive solutions for water resource management.

## 2. Methodology and Materials

### 2.1. Data Sources

The literature data adopted for this research were acquired from two common and influential scientific databases, i.e., the Science Citation Index Expanded (SCI–E) and the Social Science Citation Index (SSCI) of WOS [11]. The following terms were used to retrieve related publications: TS = ("uncertain*" AND "water" AND ("modeling" OR "simulat*") AND ("basin" OR "watershed") AND ("manage*" OR "allocation")) ("TS" represents an article subject and "*" represents a fuzzy search). Under these conditions, a total of 2020 documents with full bibliographic records were retrieved and downloaded as related research from 1991 to 2018.

### 2.2. Statistical Methods

In the realm of bibliometric studies, CiteSpace has emerged as a widely utilized tool for visualizing frontier knowledge and constructing networks within scientific domains [29]. This tool employs view maps that translate complex data into visual representations, with nodes representing various entities such as keywords, authors, journals, and countries, while links denote co–citation structures. Each node is characterized by three types of colors and different thicknesses, signifying its centrality value within the network [33,34]. For instance, a red ring around a node signifies a burst discovery, while a purple rim indicates high betweenness centrality ($\geq$0.1), which represents the significance of the node in the overall network structure [35–38].

A bibliometric study was conducted using CiteSpace, focusing on publications from 1991 to 2018 and leveraging data from the 50 most–cited journals to construct an initial knowledge network. Afterwards, each network was generated and enclosed 2020 references. The time horizon from 1991 to 2018 was divided into three periods (i.e., 1991–1999, 2000–2009, and 2010–2018). Five types involving author, institution, country, keyword, and cited reference were aligned with the research requirements, and a few default settings were maintained. Subsequently, collaboration and co–occurrence networks were systematically analyzed, considering factors such as frequency, burst, and centrality. This analytical approach aimed to identify and characterize research trends and patterns in uncertainty modeling and management within watershed studies. Through these analyses, the study sought to uncover the dynamics of collaboration among authors and institutions, highlight key research themes through co–occurrence networks, and ascertain the significance of specific contributions through burst analysis.

## 3. Results

### 3.1. Characteristics of Publications

The analysis of 2020 publications spanning the years 1991 to 2018, focusing on uncertainties in hydrologic simulation and water resources management, reveals a development tendency within the research landscape. These publications were categorized into three document types, with articles emerging as the dominant form, constituting a substantial 95% of the total publications. Following closely were proceedings papers and reviews, securing the second and third positions, respectively. Figure 1 illustrates the distribution of publication outputs across the 28–year period. Notably, the initial publication addressing uncertainty in the hydrological system surfaced in 1992. From this point, a remarkable escalation in publication outputs unfolded, with the number of publications soaring from a modest 1 in 1991 to a significant 390 in 2018. The exponential growth was particularly conspicuous in the last five years of the study period, indicating a surge in scholarly interest and engagement with the subject matter. This result thoroughly explains why research on uncertainty in hydrological modeling and management has attracted increasing attention from scholars.

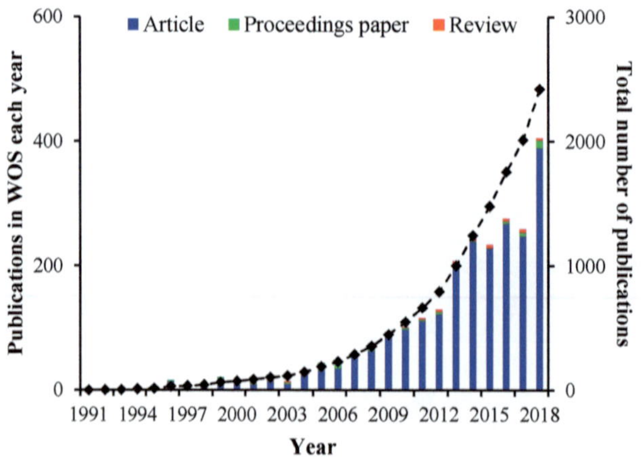


**Figure 1.** The characteristics of publications about uncertainties in hydrologic simulation and water resources management during 1991–2018.

### 3.2. *Journal Co–Citation Analysis*

Table 1 offers a comprehensive overview of the top ten frequently cited journals in the relevant domain, shedding light on the prominent sources shaping research on uncertainty in hydrological systems. The preeminent position is secured by the Journal of Hydrology, boasting an impressive 1430 publications, underscoring its pivotal role in advancing hydrological research. Following closely, Water Resources Research claims the second spot with 1368 publication. The dominance of these two journals signifies their pivotal role as major conduits for disseminating knowledge in this critical field. For the Impact Factor (IF) level, Water Resources Research and Hydrology and Earth System Sciences emerge as influential players with notably higher IF values (4.361 and 4.256, respectively). These robust IF values not only reflect the journals' academic rigor but also signify their significant impact within the broader realm of hydrological research. The higher IF values denote a more substantial influence in the related areas, further emphasizing the pivotal role that these journals play in shaping the discourse on uncertainty in hydrological systems. Researchers and practitioners in the field can leverage these insights to navigate the wealth of literature and stay informed about the latest developments, methodologies, and advancements in hydrological uncertainty analysis.

**Table 1.** Top10 journals according to frequency.

| Journal | Frequency | Centrality | IF (2017) |
|---|---|---|---|
| Journal of Hydrology | 1430 | 0.18 | 3.73 |
| Water Resources Research | 1368 | 0.29 | 4.36 |
| Hydrological Processes | 963 | 0.10 | 3.18 |
| Journal of the American Water Resources Association | 777 | 0.09 | 2.16 |
| Water Resources Management | 773 | 0.05 | 2.64 |
| Hydrology and Earth System Sciences | 744 | 0.03 | 4.26 |
| Environmental Modelling Software | 715 | 0.01 | 4.18 |
| Advances in Water Resources | 552 | 0.08 | 3.51 |
| Hydrological Sciences Journal | 547 | 0.10 | 2.06 |
| Journal of Environmental Management | 499 | 0.10 | 4.01 |

### 3.3. Country/Territory and Institution Cooperation Analysis

The global spotlight on research addressing the uncertainties in hydrologic simulation and water resources management is evident in the attention garnered by this field. Detailed insights into the top ten productive countries/territories and institutions are provided in Table 2, shedding light on the geographical distribution of research contributions. The results highlight the USA as the foremost contributor, hosting the maximum number of journals dedicated to the field. Following closely are China, Canada, and Australia, with China standing out as the only developing country among the selected nations and territories. Notably, the cumulative output from the top 10 organizations constitutes a significant portion, contributing to 29.2% of the total publications. Delving into institutional contributions, the Chinese Academy of Sciences emerges as the frontrunner, making a substantial impact with 113 publications. Following closely are Beijing Normal University (China) and the University of Regina (Canada), thereby further exemplifying the global collaborative nature inherent in research on hydrologic simulation and water resource management. These findings underscore the diverse and impactful efforts made by institutions across the globe, with each playing a crucial role in advancing our understanding of uncertainties in hydrology and water resource systems.

**Table 2.** Top 10 most productive countries based on total publications during 1991–2018.

| Rank | Country/Region | Number | Institution | Number |
|---|---|---|---|---|
| 1 | USA | 759 | Chinese Acad Sci | 113 |
| 2 | China | 507 | Beijing Normal Univ | 84 |
| 3 | Canada | 220 | Univ Regina | 80 |
| 4 | Australia | 139 | North China Elect Power Univ | 68 |
| 5 | Germany | 123 | Texas A&M Univ | 50 |
| 6 | The Netherlands | 101 | USDA ARS | 45 |
| 7 | UK | 100 | China Agr Univ | 43 |
| 8 | Iran | 79 | Peking Univ | 40 |
| 9 | Spain | 70 | US Geol Survey | 34 |
| 10 | Italy | 66 | Delft Univ Technol | 33 |

To obtain more collaboration information about countries and institutions, a detailed presentation of the first 20 co–country/territory and co–institution networks are presented in detail in Figure 2. The United States and China emerge as pivotal players in global cooperation, exerting significant influence, particularly on countries like Canada and Australia. Noteworthy relationships are highlighted, such as the impactful collaboration between China and Canada, underscoring the interconnectedness of research efforts on a global scale. And the symbiotic ties between the Chinese Academy of Sciences and Beijing Normal University underscore the importance of institutional collaboration. Additionally, the analysis also illustrates areas for potential enhancement in international influence. For instance, India and Tsinghua University are identified as entities that could amplify their impact either by increasing their publication output or fostering closer collaborations in the relevant domain. This insight serves as valuable guidance for countries and institutions seeking to strengthen their global presence and contribute meaningfully to advancements in hydrological research and water resource management. As the interconnected landscape of global collaboration unfolds, fostering strategic partnerships and bolstering international influence emerge as key considerations for the continued progress and innovation in the field.

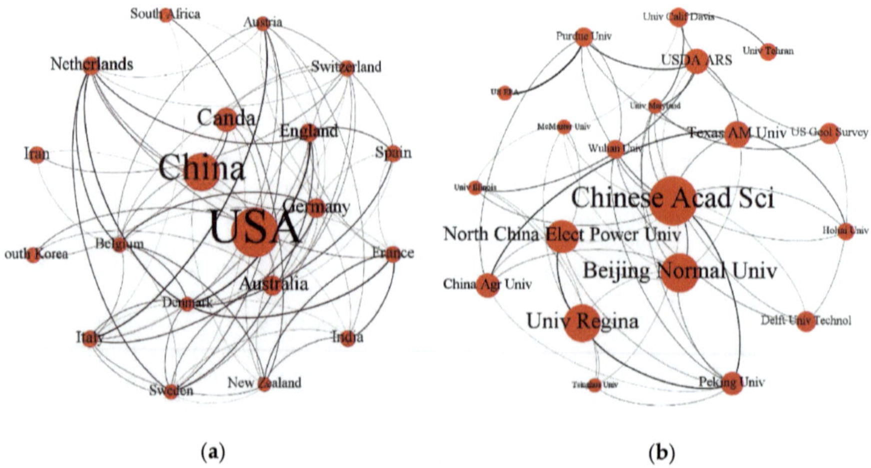


**Figure 2.** The cooperation network of the top 20 countries/territories and institutions: (**a**) network map of countries/territories; (**b**) network map of institutions.

### 3.4. Author Co–Citation Analysis

In Figure 3, an item density visualization unveils the most influential authors shaping the landscape of uncertainty in hydrological systems. Notably, the preeminent figure in hydrological modeling is Arnold from the United States Department of Agriculture, whose impactful contributions have garnered widespread recognition, reflected in an impressive citation frequency of 354 [38]. Arnold's seminal research serves as a cornerstone in the field, evidenced by its broad incorporation and acknowledgment by peers. Following closely is Beven from Lancaster University, England, occupying the second position as an influential researcher. Subsequent noteworthy contributors include Wilby from the University of Derby, England, Huang from the University of Regina, Canada, and Bergstrom from the Karolinska Institution, Sweden. These scholars, prominently featured in the item density visualization, have collectively played pivotal roles in advancing our understanding of uncertainty in hydrological systems. It is discernible from the results that these influential figures share interconnected research threads, underscoring a collaboration in their significant contributions to the field.

### 3.5. Reference Citation Bursts Analysis

The identification of citation bursts in a publication is a strong indicator of its widespread recognition within its scientific domain. Moreover, it serves as a tool for discovering emerging research fields, often characterized by clusters with citation bursts [39]. In Table 3, where the top ten references are presented, the time horizon spans from 1991 to 2018, with the red line denoting the citation bursts. Notably, the first significant milestone paper in this field, dating back to 1991, focused on model evaluation in simulations and exhibited a substantial burst strength of 24.61 [10]. Following closely, the second highest burst–ranking reference delves into the calibration and validation of the SWAT model, boasting a burst strength of 18.6 [40]. Remarkably, eight references within this study are closely linked to the SWAT model, suggesting its prominence as possibly the most widely adopted model in hydrological modeling and water resources management. The visual representation of citation bursts not only highlights pivotal contributions but also offers valuable insights into the temporal dynamics and evolving emphases within this research domain.

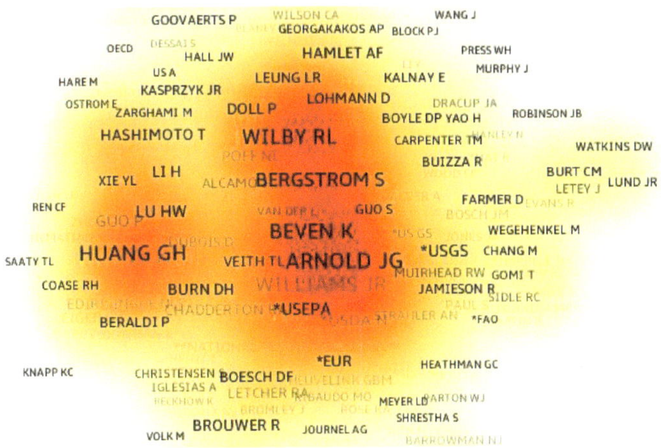


**Figure 3.** Main co–cited authors contributing to publications (* and ** represent agencies. *USEPA: United States Environmental Protection Agency; *EUR: European Research Council; *USGS: European Research Council; National Research Council: National Research Council of United States).

**Table 3.** Top 10 references with the strongest citation bursts (Red bars: Indicate periods of citation bursts where a reference is cited significantly more frequently compared to other times. Blue or light blue bars: The citation frequency of the reference is at a normal or baseline level, without significant peaks).

| References | Strength | Begin | End | 1991–2018 |
|---|---|---|---|---|
| Moriasi et al., 2007 [10] | 24.61 | 2013 | 2015 | |
| Arnold et al., 2012 [40] | 18.60 | 2015 | 2018 | |
| Abbaspour et al., 2015 [41] | 14.80 | 2016 | 2018 | |
| Abbaspour et al., 2007 [42] | 13.88 | 2013 | 2015 | |
| Yang et al., 2008 [43] | 12.72 | 2013 | 2016 | |
| Gassman et al., 2007 [44] | 12.64 | 2013 | 2015 | |
| Taylor et al., 2012 [45] | 11.29 | 2015 | 2018 | |
| Beven et al., 2001 [14] | 8.61 | 2004 | 2009 | |
| Huang et al., 2012 [46] | 7.98 | 2016 | 2018 | |
| Harou et al., 2009 [47] | 7.94 | 2015 | 2018 | |

### 3.6. Subject Categories Co–Occurrence Analysis

The dual–map overlays offer a comprehensive view of the intricate landscape of scientific journals, incorporating data from over 10,000 journals sourced from WOS [48]. In Figure 4, the visualization of publications spanning the years 1991 to 2018 pertaining to the topic of uncertainty in hydrological systems is presented, revealing a dynamic and multifaceted network. The colored curves in the figure delineate the process from left to right, showcasing the evolution of research fields and their interconnections. The distinct separation of citing and cited maps across various research fields is evident in the visualization. Each color cluster is indicative of a specific field, providing a visual representation of the interdisciplinary nature of uncertainty in hydrological systems research. Notably, three primary domains stand out in the landscape of publications on uncertainty in hydrologic modeling and water resource management: the blue cluster corresponds to Ecology/Earth/Marine, the yellow to Veterinary/Animal/Science, and the red to Mathematics/Systems/Mathematical. The blue cluster forms a significant portion of publications and citation links, suggesting a robust foundation in environmental science. Intriguingly, the yellow cluster demonstrates a unique intersection between hydrological systems and veterinary sciences. Meanwhile, the red cluster highlights the quantitative and mathematical aspects inherent in uncertainty analyses. Furthermore, the spatial distribution of citation links originating from the upper right corner of the map indicates a broad interdisciplinary approach. Publications on the uncertainty in hydrological systems draw from diverse disciplines, including environmental science, ecology, geology, mathematics, and chemistry. This interdisciplinary nature underscores the complex and multifaceted character of uncertainty in hydrological modeling and water resource management, emphasizing the need for a holistic and integrated approach to address the challenges posed by uncertainties in this critical field of study.

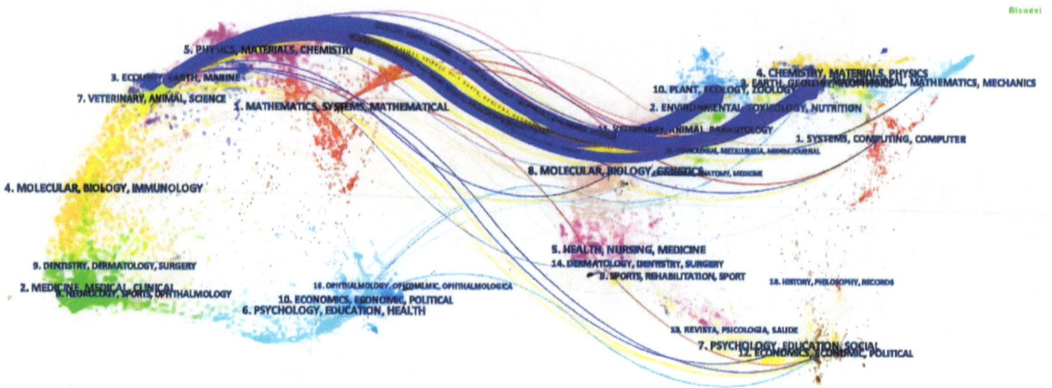


**Figure 4.** A dual–map overlay of the science mapping literature between 1991 and 2018.

### 3.7. Keywords Co–Word Analysis

The co–word study performed in this research serves a crucial function in uncovering hotspots and understanding the structure of the relevant scientific domain [36,49]. To construct the co–occurrence network (Figure 5), the first 20 items were selected, each represented by a cross, where the size reflects its frequency—larger sizes indicating higher occurrence rates. In this Figure, "Uncertainty" emerges as the most frequent keyword with 932 occurrences, followed by "basin" (584), "model" (492), "management" (447), and "climate change" (437). "Uncertainty" emerges as the most frequent keyword with 932 occurrences, followed by "basin" (584), "model" (492), "management" (447), and "climate change" (437). The centrality of each node in the network signifies its importance, and among the top 20 keywords, "water quality", "management", and "simulation" stand out with high centrality values, indicating their critical roles in the field of uncertainty in hydrological systems. This underscores the heightened attention towards research on uncertainties associated with water quality, reflecting a significant focus within the broader scope of hydrological modeling and water resources management.

*water*

MDPI

*Review*

# A Scientometric Review for Uncertainties in Integrated Simulation–Optimization Modeling System

**Congcong Li [1,2], Lulu He [3,*], Dan Liu [1] and Zhiyong Feng [2,4]**

[1] Department of Hydraulic Engineering, Tsinghua University, Beijing 100084, China; ccli@mail.bnu.edu.cn (C.L.)
[2] Key Laboratory of Water Grid Project and Regulation of Ministry of Water Resources, Wuhan 430014, China
[3] School of Mechanics and Civil Engineering, China University of Mining and Technology (Beijing), Beijing 100083, China
[4] Changjiang Institute of Survey, Planning, Design and Research, Wuhan 430014, China
* Correspondence: luckiesthe@163.com

**Highlights:**

- Conducted a comprehensive review analysis of over 2000 articles in integrated simulation–optimization modeling systems, revealing significant advancements in hydrologic modeling and water resource management.
- Unveiled the knowledge structure, frontiers, influential regions, scholars, and publications in the field using advanced visualization techniques.
- Integrated GIS, environmental science, and data science to present a multidimensional perspective on water resource management.
- Highlighted the impact of climate change on water resource management, offering adaptive management methods and contributing to policy making, guiding future research directions and practical applications.

**Abstract:** Water resources management is a challenging task caused by huge uncertainties and complexities in hydrological processes and human activities. Over the last three decades, various scholars have carried out the study on hydrological simulation under complex conditions and quantitatively characterized the associated uncertainties for water resources systems. To keep abreast of the development of the collective knowledge in this field, a scientometric review and metasynthesis of the existing uncertainty analysis research for supporting hydrological modeling and water resources management has been conducted. A total of 2020 publications from 1991 to 2018 were acquired from the Web of Science. The scientific structure, cooperation, and frontiers of the related domain were explored using the science mapping software CiteSpace V5.4.R3. Through co-citation, collaboration, and co-occurrence network study, the results present the leading contributors among all countries and hotspots in the research domain. In addition, synthetical uncertainty management for hydrological models and water resource systems under climatic and land use change will continue to be focused on. This study comprehensively evaluates various aspects of uncertainty analysis in hydrologic simulation–optimization systems, showcasing advanced data analysis and artificial intelligence technologies. It focuses on current research frontiers, aiding decision-makers in better understanding and managing the complexity and uncertainties of water resource systems, thereby enhancing the sustainability and efficiency of responses to environmental changes.

**Keywords:** uncertainty; hydrologic modeling; water resources management; visualization analysis; CiteSpace; review

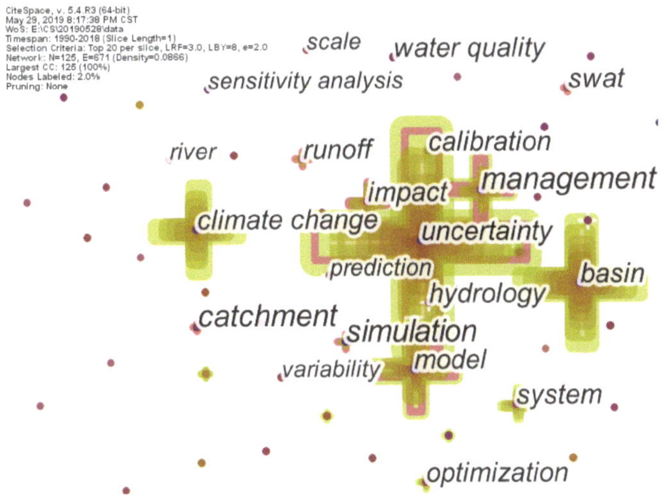


**Figure 5.** Keyword co–occurrence map of publications.

### *3.8. Document Co-Citation Analysis*

#### 3.8.1. Research Cluster Analysis

The reference relationships within the realm of uncertainty in hydrologic modeling and water resources management not only adhere to the objective law of scientific development but also intricately reflect the intellectual structure of the field [12]. To further determine the distribution rule of references, the top 20 references between 1991 and 2018 were picked out. Figure 6 vividly illustrates the central clusters that define the domain of uncertainty in hydrologic modeling and water resources management. The high modularity value of 0.76 underscores a distinctly defined landscape within the field, providing a clear conceptual framework [38,39,50,51]. Within these references, a granular dissection reveals 96 clusters, with 14 of them appropriately labeled. The study field on uncertainty in hydrologic modeling and water resources management exhibits a multifaceted nature, encompassing diverse aspects such as management objects (water resource and water quality), measures (stochastic programming and statistical learning), simulation techniques (SWAT), technological applications (GIS), and specific study areas (China and Canada). This comprehensive coverage suggests that the study of uncertainty in hydrological modeling and integrated management has evolved into a relatively mature and nuanced area of research in recent years. The multitude of identified clusters and labeled aspects attests to the richness and depth of scholarship within this domain.

#### 3.8.2. Timeline View of Typical Clusters

The timeline visualization presented in Figure 7 offers a comprehensive overview of the temporal evolution of various clusters, providing valuable insights into the trajectory and longevity of distinct research themes. Ten prominent clusters are delineated along horizontal timelines, each encapsulating a unique set of studies and developments. Figure 7 not only portrays the temporal progress of these clusters but also highlights the three most cited references associated with each timeline. Examining the illustration, it becomes evident that individual clusters exhibit varying temporal spans, with some enduring for approximately 20 years, while others have a more concise duration. For instance, clusters such as #0 and #6, focusing on the SWAT model, demonstrate a prolonged period of significance from 1994 to 2016. Essential achievements within these clusters are notably concentrated between 2005 and 2015, showcasing a concentrated period of impactful research.

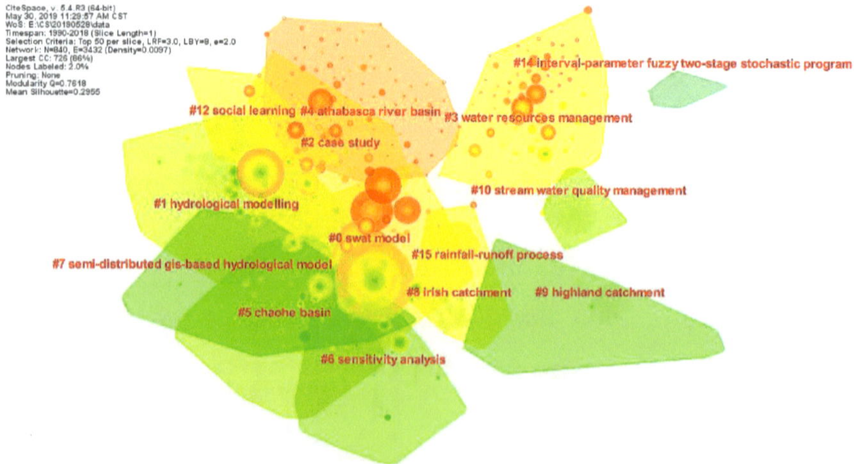


**Figure 6.** A cluster view of the co–citation network.

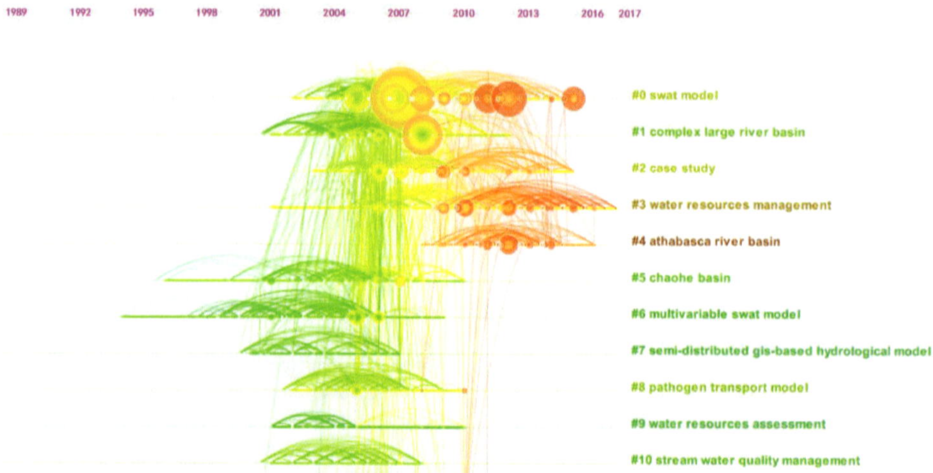


**Figure 7.** A timeline visualization of the top 10 study clusters.

Cluster #4, centered on uncertain management, stands out with a commendable span of 16 years, underscoring its enduring relevance and continued activity. In contrast, cluster #7, centered on the GIS–based model, concludes in 2007, suggesting a shift in research specialties and the emergence of new directions within the relevant research landscape. The timeline visualization not only serves as a historical record of the longevity of research clusters but also allows for the identification of pivotal periods and transformative shifts in research focus. It provides a dynamic perspective on the evolution of key themes, enabling researchers to trace the development and impact of various clusters over time.

## 4. Conclusions

This research offers a comprehensive scientometric review that synthesizes the state of uncertainty analysis in hydrologic simulation and water resources systems through a meticulous examination of bibliographic data. The investigation encompasses diverse dimensions, including the characteristics of publications, collaboration among countries/territories and institutions, co–citation of scholars and references, as well as the co–occurrence of topics and keywords. By delving into these multifaceted aspects, the study provides valuable insights into the current state of research in this critical domain.

Over the past three decades, there have been approximately 2020 publications related to the uncertainty in the hydrological system, and the number has increased steadily, which indicates that this field is receiving increasing attention from scholars. Generally, the USA, China, Canada, Australia, and Germany were the first five prolific countries, and at the same time, the Chinese Academy of Sciences, Beijing Normal University, University of Regina, North China Electric Power University, and Texas A&M University were the first five prolific institutions in this related field. The USA contributed greatly to the publications and cooperated with most countries/territories. The Chinese Academy of Sciences was the leading institution and had a relatively intimate relationship with other institutions.

Most studies were published in the representative journals in this field, such as the Journal of Hydrology and Water Resources Research. Arnold, Beven, and Huang were the representative scholars who made outstanding contributions to the field of uncertainty in hydrologic simulation and water resources management. Research on the uncertainty in hydrologic simulation and water resources management covered broad subjects, such as environmental science, ecology, geology, mathematics, and chemistry. Through the keyword analysis, uncertainty, hydrological simulation, climate change, calibration, and optimization were the hotspots in the research domain. Research on water quality and sensitivity analysis have been proven new topics over the years. By evaluating the co–cited network, the primary study field was determined to be hydrological simulations using the SWAT model, and this topic is becoming more and more mature.

In conclusion, this study field for the uncertainty in hydrologic simulation and water resources management is still a research front, and it needs to be further explored and improved in the following aspects. (i) The current uncertainty research mainly concentrates on hydrological models, and further exploration in areas such as structure, parameters, and data collection can enhance the reliability of hydrological predictions and decision–making. (ii) The uncertainty analysis of hydrological systems is insufficient to study the water–transforming pattern within atmospheric water, surface water, and groundwater included in the hydrological cycle, and new methods should be used to explore each link of the eco–hydrological process. (iii) As a single method was unable to meet the study of the uncertainty in the hydrological system, future research will likely focus on the establishment of coupled uncertainty analysis methods for synthetical uncertainty in hydrological model application and management. This research provides a broad perspective for uncertainties in integrated simulation–optimization modeling system. Furthermore, the study contributes to the development of more informed and effective strategies for water system management, especially in addressing challenges such as water scarcity, land use, and climate change. In summary, this research not only advances our understanding of hydrological processes but also provides sustainable adaptive solutions for water resource management.

**Funding:** This study was supported by the Key Lab of Water Grid Project and Regulation of Ministry of Water Resources (QTKS0034W23288), the National Natural Science Foundation of China (42307558, 52300246), and the National Key Research Program of China (2022YFE0101100).

**Data Availability Statement:** Some or all data, models, or code that support the findings of this study are available from the corresponding author upon reasonable request.

**Acknowledgments:** The authors would like to express gratitude to the journal editor and anonymous reviewers for their comments and suggestions.

**Conflicts of Interest:** The authors declare no conflicts of interest.

## References


1. Akter, A.; Babel, M.S. Hydrological modeling of the Mun River basin in Thailand. *J. Hydrol.* **2012**, *452*, 232–246. [CrossRef]
2. Cho, K.H.; Pachepsky, Y.A.; Kim, M.; Pyo, J.; Park, M.H.; Kim, Y.M.; Kim, J.W.; Kim, J.H. Modeling seasonal variability of fecal coliform in natural surface waters using the modified SWAT. *J. Hydrol.* **2016**, *535*, 377–385. [CrossRef]
3. Pereira, D.D.; Martinez, M.A.; Pruski, F.F.; da Silva, D.D. Hydrological simulation in a basin of typical tropical climate and soil using the SWAT model part I: Calibration and validation tests. *J. Hydrol.* **2016**, *7*, 14–37. [CrossRef]
4. Wang, W.K.; Dai, Z.X.; Zhao, Y.Q.; Li, J.T.; Duan, L.; Wang, Z.F.; Zhu, L. A quantitative analysis of hydraulic interaction processes in stream aquifer systems. *Sci. Rep.* **2016**, *6*, 19876. [CrossRef]
5. Xue, L.Q.; Yang, F.; Yang, C.B.; Wei, G.H.; Li, W.Q.; He, X.L. Hydrological simulation and uncertainty analysis using the improved TOPMODEL in the arid Manas River basin, China. *Sci. Rep.* **2018**, *8*, 452. [CrossRef]
6. Fonseca, A.; Botelho, C.; Boaventura, R.A.R.; Vilar, V.J.P. Integrated hydrological and water quality model for river management: A case study on Lena River. *Sci. Total Environ.* **2014**, *485*, 474–489. [CrossRef]
7. Ma, Y.D.; Zhao, J.B.; Luo, X.Q.; Shao, T.J.; Dong, Z.B.; Zhou, Q. Hydrological cycle and water balance estimates for the megadune-lake region of the Badain Jaran Desert, China. *Hydrol. Process.* **2017**, *31*, 3255–3268. [CrossRef]
8. Xia, J.; Zhang, Y.Y.; Zhan, C.S.; Ye, A.Z. Water Quality Management in China: The Case of the Huai River Basin. *Int. J. Water Res. Dev.* **2011**, *27*, 167–180. [CrossRef]
9. Montanari, A.; Brath, A. A stochastic approach for assessing the uncertainty of rainfall-runoff simulations. *Water Resour. Res.* **2004**, *40*, W01106. [CrossRef]
10. Moriasi, D.N.; Arnold, J.G.; Van Liew, M.W.; Bingner, R.L.; Harmel, R.D.; Veith, T.L. Model evaluation guidelines for systematic quantification of accuracy in watershed simulations. *Trans. ASABE* **2007**, *50*, 885–900. [CrossRef]
11. Wu, X.L.; Chen, X.Y.; Zhan, F.B.; Hong, S. Global research trends in landslides during 1991–2014: A bibliometric analysis. *Landslides* **2015**, *12*, 1215–1226. [CrossRef]
12. Yin, L.C.; Kretschmer, H.; Hanneman, R.A.; Liu, Z.Y. Connection and stratification in research collaboration: An analysis of the COLLNET network. *Inf. Process. Manag.* **2006**, *42*, 1599–1613. [CrossRef]
13. Beven, K.; Binley, A. The future of distributed models—Model calibration and uncertainty prediction. *Hydrol. Process.* **1992**, *6*, 279–298. [CrossRef]
14. Beven, K.; Freer, J. Equifinality, data assimilation, and uncertainty estimation in mechanistic modelling of complex environmental systems using the GLUE methodology. *J. Hydrol.* **2001**, *249*, 11–29. [CrossRef]
15. Krzysztofowicz, R. Bayesian theory of probabilistic forecasting via deterministic hydrologic model. *Water Resour. Res.* **1999**, *35*, 2739–2750. [CrossRef]
16. Blasone, R.S.; Madsen, H.; Rosbjerg, D. Uncertainty assessment of integrated distributed hydrological models using GLUE with Markov chain Monte Carlo sampling. *J. Hydrol.* **2008**, *353*, 18–32. [CrossRef]
17. Bouda, M.; Rousseau, A.N.; Konan, B.; Gagnon, P.; Gumiere, S.J. Bayesian Uncertainty Analysis of the Distributed Hydrological Model HYDROTEL. *J. Hydrol. Eng.* **2012**, *17*, 1021–1032. [CrossRef]
18. Mantovan, P.; Todini, E. Hydrological forecasting uncertainty assessment: Incoherence of the GLUE methodology. *J. Hydrol.* **2006**, *330*, 368–381. [CrossRef]
19. Vazquez, R.F.; Beven, K.; Feyen, J. GLUE Based Assessment on the Overall Predictions of a MIKE SHE Application. *Water Resour. Manag.* **2009**, *23*, 1325–1349. [CrossRef]
20. Du, P.; Li, Y.P.; Huang, G.H. Inexact Chance-Constrained Waste-Load Allocation Model for Water Quality Management of Xiangxihe River. *J. Environ. Eng.* **2013**, *139*, 1178–1197. [CrossRef]
21. Li, Y.P.; Huang, G.H.; Huang, Y.F.; Zhou, H.D. A multistage fuzzy-stochastic programming model for supporting sustainable water-resources allocation and management. *Environ. Model. Softw.* **2009**, *24*, 786–797. [CrossRef]
22. Huang, G.H.; Loucks, D.P. An inexact two-stage stochastic programming model for water resources management under uncertainty. *Civ. Eng. Environ. Syst.* **2000**, *17*, 95–118. [CrossRef]
23. Guo, P.; Huang, G.H.; Li, Y.P. Inexact Fuzzy-Stochastic Programming for Water Resources Management Under Multiple Uncertainties. *Environ. Model. Assess.* **2010**, *15*, 111–124. [CrossRef]
24. Li, Y.P.; Huang, G.H.; Zhang, N.; Nie, S.L. An inexact-stochastic with recourse model for developing regional economic-ecological sustainability under uncertainty. *Ecol. Model.* **2011**, *222*, 370–379. [CrossRef]
25. Li, Y.P.; Liu, J.; Huang, G.H. A hybrid fuzzy-stochastic programming method for water trading within an agricultural system. *Agric. Sys.* **2014**, *123*, 71–83. [CrossRef]
26. Li, C.C.; Cai, Y.P.; Qian, J.P. A multi-stage fuzzy stochastic programming method for water resources management with the consideration of ecological water demand. *Ecol. Indic.* **2018**, *95*, 930–938. [CrossRef]
27. Zhuang, X.W.; Li, Y.P.; Nie, S.; Fan, Y.R.; Huang, G.H. Analyzing climate change impacts on water resources under uncertainty using an integrated simulation-optimization approach. *J. Hydrol.* **2018**, *556*, 523–538. [CrossRef]
28. Li, C.C.; Cai, Y.P.; Tan, Q.; Wang, X.; Li, C.H.; Liu, Q.; Chen, D.N. An integrated simulation-optimization modeling system for water resources management under coupled impacts of climate and landuse variabilities with priority in ecological protection. *Adv. Water Resour.* **2021**, *154*, 103986. [CrossRef]
29. Chen, C.M. Searching for intellectual turning points: Progressive knowledge domain visualization. *Proc. Natl. Acad. Sci. USA* **2004**, *101*, 5303–5310. [CrossRef]

30. Garcia-Lillo, F.; Ubeda-Garcia, M.; Marco-Lajara, B. Organizational ambidexterity: Exploring the knowledge base. *Scientometrics* **2016**, *107*, 1021–1040. [CrossRef]
31. Merigo, J.M.; Yang, J.B. A bibliometric analysis of operations research and management science. *Omega* **2017**, *73*, 37–48. [CrossRef]
32. Wang, H.; Xu, Z.S. Admissible orders of typical hesitant fuzzy elements and their application in ordered information fusion in multi-criteria decision making. *Inf. Fusion* **2016**, *29*, 98–104. [CrossRef]
33. Chen, C.M. CiteSpace II: Detecting and visualizing emerging trends and transient patterns in scientific literature. *J. Am. Soc. Inf. Sci. Technol.* **2006**, *57*, 359–377. [CrossRef]
34. Chen, C.M.; Ibekwe-SanJuan, F.; Hou, J.H. The Structure and Dynamics of Cocitation Clusters: A Multiple-Perspective Cocitation Analysis. *J. Am. Soc. Inf. Sci. Technol.* **2010**, *61*, 1386–1409. [CrossRef]
35. Xie, P. Study of international anticancer research trends via co-word and document co-citation visualization analysis. *Scientometrics* **2015**, *105*, 611–622. [CrossRef]
36. Freeman, L.C. Centrality in Social Networks Conceptual Clarification. *Soc. Netwks.* **1979**, *1*, 215–239. [CrossRef]
37. Liu, Z.G.; Yin, Y.M.; Liu, W.D.; Dunford, M. Visualizing the intellectual structure and evolution of innovation systems research: A bibliometric analysis. *Scientometrics* **2015**, *103*, 135–158. [CrossRef]
38. Arnold, J.G.; Srinivasan, R.; Muttiah, R.S.; Williams, J.R. Large area hydrologic modeling and assessment—Part 1: Model development. *JAWRA J. Am. Water Resour. Assoc.* **1998**, *34*, 73–89. [CrossRef]
39. Chen, C.M.; Dubin, R.; Kim, M.C. Emerging trends and new developments in regenerative medicine: A scientometric update (2000–2014). *Expert Opin. Biol. Ther.* **2014**, *14*, 1295–1317. [CrossRef]
40. Yu, D.J. A scientometrics review on aggregation operator research. *Scientometrics* **2015**, *105*, 115–133. [CrossRef]
41. Arnold, J.G.; Moriasi, D.N.; Gassman, P.W.; Abbaspour, K.C.; White, M.J.; Srinivasan, R.; Santhi, C.; Harmel, R.D.; van Griensven, A.; Van Liew, M.W.; et al. SWAT: Model use, calibration, and validation. *Trans. ASABE* **2012**, *55*, 1491–1508. [CrossRef]
42. Abbaspour, K.C.; Rouholahnejad, E.; Vaghefi, S.; Srinivasan, R.; Yang, H.; Klove, B. A continental-scale hydrology and water quality model for Europe: Calibration and uncertainty of a high-resolution large-scale SWAT model. *J. Hydrol.* **2015**, *524*, 733–752. [CrossRef]
43. Abbaspour, K.C.; Yang, J.; Maximov, I.; Siber, R.; Bogner, K.; Mieleitner, J.; Zobrist, J.; Srinivasan, R. Modelling hydrology and water quality in the pre-alpine/alpine Thur watershed using SWAT. *J. Hydrol.* **2007**, *333*, 413–430. [CrossRef]
44. Yang, J.; Reichert, P.; Abbaspour, K.C.; Xia, J.; Yang, H. Comparing uncertainty analysis techniques for a SWAT application to the Chaohe Basin in China. *J. Hydrol.* **2008**, *358*, 1–23. [CrossRef]
45. Gassman, P.W.; Reyes, M.R.; Green, C.H.; Arnold, J.G. The soil and water assessment tool: Historical development, applications, and future research directions. *Trans. ASABE* **2007**, *50*, 1211–1250. [CrossRef]
46. Taylor, K.E.; Stouffer, R.J.; Meehl, G.A. AN Overview of Cmip5 and The Experiment Design. *Bull. Am. Meteorol. Soc.* **2012**, *93*, 485–498. [CrossRef]
47. Huang, Y.; Li, Y.P.; Chen, X.; Ma, Y.G. Optimization of the irrigation water resources for agricultural sustainability in Tarim River Basin, China. *Agric. Water Manag.* **2012**, *107*, 74–85. [CrossRef]
48. Harou, J.J.; Pulido-Velazquez, M.; Rosenberg, D.E.; Medellin-Azuara, J.; Lund, J.R.; Howitt, R.E. Hydro-economic models: Concepts, design, applications, and future prospects. *J. Hydrol.* **2009**, *375*, 627–643. [CrossRef]
49. Chen, C.M.; Leydesdorff, L. Patterns of Connections and Movements in Dual-Map Overlays: A New Method of Publication Portfolio Analysis. *J. Assoc. Inf. Sci. Technol.* **2014**, *65*, 334–351. [CrossRef]
50. Xiang, C.Y.; Wang, Y.; Liu, H.W. A scientometrics review on nonpoint source pollution research. *Ecol. Eng.* **2017**, *99*, 400–408. [CrossRef]
51. Chen, C.M. Science Mapping: A Systematic Review of the Literature. *J. Data Inf. Sci.* **2017**, *2*, 1–40. [CrossRef]

*water*

MDPI

*Article*

# Joint Optimal Use of Sluices of a Group of Cascade Hydropower Stations under High-Intensity Peak Shaving and Frequency Regulation

**Shiyu Mou [1], Tian Qu [1], Jia Li [1], Xin Wen [2,*] and Yu Liu [2]**

[1] Dadu River Basin Production Command Center, China Energy Investment, Chengdu 610041, China; sy18014399579@163.com (S.M.); qutianddh@163.com (T.Q.); lijiaddh@163.com (J.L.)
[2] College of Water Conservancy and Hydropower Engineering, Hohai University, Nanjing 210098, China; nz15982630250@163.com
* Correspondence: ly15261466541@163.com

**Abstract:** With the large-scale development and grid connection of renewable energy, hydropower faces more intense and frequent peak shaving and frequency regulation, giving rise to water level fluctuations and frequently forced sluice adjustments at hydropower stations. This paper proposes a model that combines "offline calculation" and "online search". First, feasible sluice opening combinations for different water levels at each hydropower station are calculated offline, and a sluice operation strategy table is constructed. Subsequently, an optimal sluice operation strategy is searched online according to the real-time water level and various regulatory requirements. As an example, we select three hydropower stations in the middle reach of the Dadu River in China, namely, Pubugou, Shenxigou, and Zhentouba. The results show that the total number of adjustments of the sluices of the cascade hydropower stations was reduced from 1195 to 675, a reduction of 43.5%, and the leading hydropower station, Pubugou, met water level control requirements, whereas the fluctuations in the water level of the two downstream daily regulating hydropower stations, Shenxigou and Zhentouba, were reduced by 1.38 m and 0.55 m, respectively. The results indicate that the sluices of hydropower stations were optimally used under high-intensity peak shaving and frequency regulation.

**Keywords:** reservoir scheduling; flood control scheduling; sluice operation; water level regulation; Dadu River Basin

## 1. Introduction

A typical cascade hydropower development is composed of a leading power station with a large storage capacity and multiple run-of-the-river power stations whose storage capacity is regulated on a daily basis [1,2]. With the recent large-scale development and connection of renewable energy to the grid, hydropower stations face more intense and frequent peak shaving and frequency regulation tasks [3–5]. For power stations subjected to daily regulation with poor storage control, frequent changes in load instructions are likely to cause large fluctuations in the water level. During the flood season, the drastic changes in water level greatly increase the risk of flood control operation in cascade reservoirs. Thus, sluices must be used to regulate reservoir levels and downstream flow to ensure the safety of reservoir dams, and downstream objects need to be protected. Traditional decision-making regarding the management of power station sluices relies on scheduling handbooks or human experience. However, high-intensity peak shaving and frequency regulation have led to more complex and repeated changes in upstream and downstream hydraulic conditions, meaning that traditional calculation methods based on water balance have difficulties complying with the accuracy requirements of water level regulation. In turn, this causes a mismatch in the flow of the upstream and downstream power stations, resulting in frequent adjustments to the operation and opening of sluices. Furthermore, small power stations are often close to each other and have close hydraulic connections. Transmission times vary from several minutes to several hours, requiring sluice operation decisions to be generated quickly. Thus, the traditional decision-making method is labor-intensive, high-risk, and insufficiently responsive. In addition, the nonlinear relationship between water level, flow rate, and sluice opening, as well as the time-varying characteristics of the sluice opening flow rate with the reservoir water level, further compound the complexity of the problem. Therefore, solving the two complex problems of accuracy and efficiency of decision-making regarding sluice operation is the key to optimizing power stations under high-intensity peak shaving and frequency regulation.

In recent years, scholars have conducted research from different perspectives, such as energy and revenue optimization, flood control scheduling, and joint operation of sluices. In terms of energy and revenue optimization, scholars have established different models and compared their optimization effects. For example, De Ladurantaye D [6] presented both a deterministic and a stochastic mathematical model to maximize the profits obtained by selling electricity produced through a cascade of dams and reservoirs in a deregulated market, and the results demonstrated the superiority of the stochastic model over the deterministic one. Avesani D [7] presented a comparative assessment of revenues provided by short-term optimizations driven by two econometric models, which may be beneficial for hydropower companies to enhance the expected revenues from storage hydropower systems, especially those characterized by large storage capacity. Regarding flood control scheduling [8–10], reservoir flood control scheduling has the characteristics of being multiconstraint, high-dimension, nonlinearity, and difficult to solve [11]. The overall approach has evolved from one or a few reservoirs securing a single flood protection object to multi-objective refined and comprehensive scheduling of a large group of reservoirs for multiple flood protection objects across a whole basin, including optimal scheduling modeling and solving, regular scheduling rule extraction, and flood risk assessments. For example, He et al. [12] proposed the improved sparrow algorithm (ISSA) combining Cauchy mutation and reverse learning strategy, which provided a new and effective way to solve the problem of flood control optimization of reservoir groups. Wang [13] proposed a multi-objective modeling and optimization method for flood control, scheduling large reservoir clusters based on different pre-defined frameworks. This effectively solved the complex optimization problem of scheduling schemes for large-scale reservoir clusters. Xu et al. [14] proposed an optimal scheduling model for flood resource utilization to derive optimal scheduling rules for the Three Gorges in flood season, considering flood risk. Huang et al. [15] proposed a multi-objective operation and risk decision-making model to provide strategies for real-time reservoir control to reduce flood risk. Zhu et al. [16] proposed the main steps that constitute optimal flood control decision-making and performed a flood control risk assessment, which showed improved reliability of flood control decisions. Nevertheless, these studies were mostly concerned with flood events, with the water level or flow rate as decision variables, and they rarely focused on the operation of power station sluices. In terms of sluice utilization, Su et al. [17] proposed an optimal scheduling model for reservoir flood control considering spillway constraints to generate a more accurate, executable flood control scheduling strategy. Kim et al. [18] proposed an optimal sluice decision-making method based on the particle swarm optimization algorithm, which can be effectively used to determine optimal sluice scheduling rules and generate sluice opening and closing strategies during the flood season. However, these studies tended to focus on individual power stations and in-depth research on the joint use of sluices in groups of cascade power stations has not yet been reported.

This paper proposes a method named the "Sluice Decision Method Based on Sluice Operation Strategy Table", which consists of "offline calculation" and "online search" to enhance the accuracy and efficiency of sluice operation in a group of cascade power stations under high-intensity peak shaving and frequency regulation. First, an "offline calculation" of the feasible sluice operation scheme at different water levels is conducted, and a sluice operation strategy for the cascade hydropower stations is developed. Then, according to the current status and scheduling requirements of the group of cascade power stations, the optimal sluice control strategy is searched online in the strategy master list. Finally, a case study is conducted in the middle-course cascade power stations of the Dadu River.

## 2. Methods

### 2.1. Model Principles and Calculation Steps

The use of sluices in cascade hydropower stations requires simultaneous decisions on the combination of sluices and the extent to which each sluice is opened. Cascade power stations often have dozens or even hundreds of different sluices; thus, numerous options exist, with a complex nonlinear relationship between the water level, flow rate, and degree of sluice opening. In addition, the variation in the sluice flow with the reservoir water level with time further exacerbates the complexity of the problem. In the practical scheduling of hydropower stations, the traditional optimization decision-making method for sluices involves manually calculating the water level process for various sluice operation schemes at each time period and selecting sluice schemes based on this calculation. However, this method is susceptible to the curse of dimensionality, leading to potential adverse effects on computational efficiency and accuracy.

To this end, in this study, we first calculated all feasible sluice operation schemes at different water levels based on the rules of the symmetrical opening of sluices of each power station and established a master list of sluice operation strategies for power stations, which was stored to reduce real-time decision-making computations. In the real-time decision-making process, based on the control objectives of fewer adjustments of sluice and water level stability, the optimal sluice operation strategy for cascade power stations was generated using an "online search" of the master list of sluice operation strategies for power stations. Usually, the opening of sluices does not change significantly over a short period of time; hence, the real-time decision-making model uses a 15-min calculation scale. The model uses power and water forecast information as boundaries, simulates and deduces the operation of each power station in the cluster of cascade power stations within 24 h, and determines whether the intra-day scheduling process of the reservoir meets the scheduling requirements based on simulation deduction results; if so, the current sluice opening is maintained, and otherwise, based on the forecast information and scheduling requirements for the future, an "online search" is conducted to find the optimal sluice control strategy. A flowchart illustrating the calculation process is presented in Figure 1.

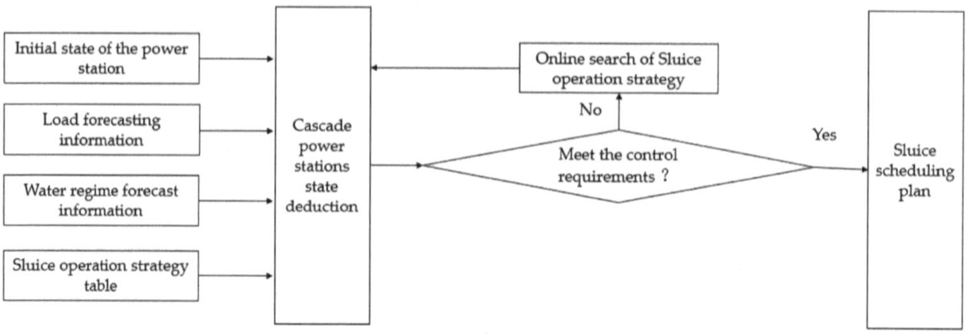


**Figure 1.** Flowchart of calculation for optimal use of sluices in a group of power stations.

### 2.2. Establishment of Sluice Operation Strategy Table Based on "Offline Calculation"

As the operation mode and requirements of the sluices at each power station are different, the feasible sluice combinations under different conditions are also different, and the decision is affected by many variables. The inclusion of all sluices in the group of cascade power stations in the optimization calculation could lead to the curse of dimensionality. Therefore, it is necessary to conduct an "offline calculation" of the feasible sluice combinations for different water levels and flow conditions, construct a sluice operation strategy table, and store the feasible combinations to achieve accurate and rational sluice regulation. We selected four indicators, namely water level, combination of sluices, sluice opening, and discharge flow rate, of each power station to construct a sluice operation strategy table for each station as follows:

Step 1: Identifying the water level interval between normal and dead levels of each power station in increments of 0.1 m.

Step 2: Determining the feasible combinations of sluice options at different water levels. Therefore, it is necessary to analyze the feasible combination of sluices at different water levels according to the sluice operation mode and operation requirements and retain the feasible combination of sluices at each water level. In the table, an open sluice is represented by 1, and a closed sluice is represented by 0.

Step 3: Enabling discrete sluice openings in increments of 0.1 m based on the feasible sluice combinations at different water levels. Considering the adverse effects of the zigzag currents on the safe operation of power stations, when multiple sluices need to be opened at each power station, it is necessary to ensure that these sluices are opened in a consistent manner.

Step 4: Calculating the sum of the flow rate of each sluice under each scheme for different water levels, sluice combinations, and sluice opening combination schemes, using the flow curve of each discharge facility. Considering the numerous scenarios covered by the sluice operation strategy table, it is impractical to display all of them. Therefore, this paper presents an example using a hydropower station with five sluices to demonstrate the sluice operation strategy table when the water level is h m, and the discharge rate is Q $m^3 \cdot s^{-1}$. Refer to Table 1 for specific details, $e_{ij}$ represents the i-th sluice in the j-th scheme.

**Table 1.** Example of sluice operation strategy table.

| Water Level/m | Discharge Rate/($m^3 \cdot s^{-1}$) | Sluice Combination | Sluice Opening/m | | | | |
|---|---|---|---|---|---|---|---|
| | | | 1# | 2# | 3# | 4# | 5# |
| h | Q | 00100 | $e_{11}$ | $e_{21}$ | $e_{31}$ | $e_{41}$ | $e_{51}$ |
| | | 01010 | $e_{12}$ | $e_{22}$ | $e_{32}$ | $e_{42}$ | $e_{52}$ |
| | | 10001 | $e_{13}$ | $e_{23}$ | $e_{33}$ | $e_{43}$ | $e_{53}$ |
| | | 10101 | $e_{14}$ | $e_{24}$ | $e_{34}$ | $e_{44}$ | $e_{54}$ |
| | | 11111 | $e_{15}$ | $e_{25}$ | $e_{35}$ | $e_{45}$ | $e_{55}$ |

### 2.3. *"Online Search"-Based Sluice Operation Strategy Generation*

After building the sluice operation strategy table, a sluice operation strategy is efficiently generated using the "online search" method. First, the operational status and processes of each cascade power station under different incoming water and load conditions are deduced based on the power requirements [19]. Based on the results of each cascade power station, it is determined whether the water level and flow rate of each station meet the control requirements, and if the control requirements are met, the extent to which the sluices are opened remains unchanged; otherwise, to meet the water level and flow constraints, an "online search" is conducted for the optimal control strategy with the objective of performing the minimum number of sluice operations.

#### 2.3.1. Objective Function

The primary optimization objective of the model is to minimize the number of sluice operations by optimizing the decision on the opening of each by optimizing the opening of each sluice. However, in calculations, situations often arise where the objective function values of multiple schemes are equal. To address such situations, the model introduces a second optimization objective function. Considering that a significant change in flood discharge flow and resulting fluctuations in water level may occur if the amplitude of two adjacent adjustments is too large, the secondary optimization objective is to minimize the opening change amplitude of adjacent sluice adjustments. Only when there are multiple solutions that result in equal values for the primary optimization objective function will the secondary objective function values of these solutions be calculated. The solution with the best secondary objective function value will then be selected among them. The objectives of the model are as follows:

$$F_1 = \min\sum_{i=1}^{n}\sum_{t=1}^{T} d_{i,t}\ ,d_{i,t} = \begin{cases} 0 & e_{i,t} = e_{i,t-1} \\ 1 & e_{i,t} \neq e_{i,t-1} \end{cases} \tag{1}$$

$$F_2 = \min\left|\sum_{i=1}^{n} e_{i,t} - \sum_{i=1}^{n} e_{i,t-1}\right| \tag{2}$$

In the formula, $i$ is the number of sluices, $i$ = 1, 2, . . ., $n$; $n$ is the total number of sluices; $t$ is the sequence number of the calculation period, $t$ = 1, 2, . . ., $T$; $T$ is the total number of time periods; $d_{i,t}$ is the number of adjustments during time period $t$ of sluice $i$ in the power stations; $e_{i,t}$ and $e_{i,t-1}$ is the opening of the sluice $i$ during the period $t$ and $t-1$.

### 2.3.2. Constraints

a. Water balance [20] constraint

$$V_{i,t+1} = V_{i,t} + (I_{i,t} - Q_{i,t})\Delta t \tag{3}$$

In the formula, $i$ is the number of the hydropower stations, $i$ = 1, 2, . . ., $N$, $N$ is the total number of cascade hydropower stations; $t$ is the sequence number of the calculation period, $t$ = 1, 2, . . ., $T$; $T$ is the total number of time periods, $T$ = 96; $V_{i,t}$ and $V_{i,t+1}$ are the reservoir storage volumes of the hydropower station $i$ at $t$ and $t+1$, respectively; $I_{i,t}$ is the average inlet flow of the hydropower station $i$ at $t$; $Q_{i,t}$ is the average outlet flow of the hydropower station $i$ at $t$; $\Delta t$ is the computational interval, $\Delta t$ = 15 min.

b. Power quantity balance constraint

$$E = \sum_{t=1}^{T}\sum_{i=1}^{N}(P_{i,t} \times \Delta t) \tag{4}$$

In the formula, $i$ is the number of the hydropower stations, $i$ = 1, 2, . . ., $N$, $N$ is the total number of cascade hydropower stations; $t$ is the sequence number of the calculation period, $t$ = 1, 2, . . ., $T$; $T$ is the total number of time periods, $T$ = 96; $P_{i,t}$ is the output power of the hydropower station $i$ in time period $t$; $E$ is the total power generation; $\Delta t$ is the computational interval, $\Delta t$ = 15 min.

c. Power balance constraint

$$P_{output,t} = \sum_{t=1}^{N} P_{i,t} \tag{5}$$

In the formula, $i$ is the number of the hydropower stations, $i$ = 1, 2, . . ., $N$, $N$ is the total number of cascade hydropower stations; $t$ is the sequence number of the calculation period, $t$ = 1, 2, . . ., $T$; $T$ is the total number of time periods, $T$ = 96; $P_{output,t}$ and $P_{i,t}$ are the total output power of cascade power stations and the hydropower station $i$ in time period $t$.

d. Power station output constraint

$$P_{i,t,\min} \leq P_{i,t} \leq P_{i,t,\max} \tag{6}$$

In the formula, $i$ is the number of the hydropower stations, $i$ = 1, 2, . . ., $N$, $N$ is the total number of cascade hydropower stations; $t$ is the sequence number of the calculation period, $t$ = 1, 2, . . ., $T$; $T$ is the total number of time periods, $T$ = 96; $P_{i,t}$ is the load of the hydropower station $i$ in time period $t$; $P_{i,t,\min}$ and $P_{i,t,\max}$ are the upper and lower limits of output of hydropower station $i$ in time period $t$.

e. Flow rate balance constraint

$$Q_{i,t} = Q_{i,t,e} + q_{i,t} \tag{7}$$

In the formula, $i$ is the number of the hydropower stations, $i = 1, 2, \ldots, N$, $N$ is the total number of cascade hydropower stations; $t$ is the sequence number of the calculation period, $t = 1, 2, \ldots, T$; $T$ is the total number of time periods, $T = 96$; $Q_{i,t}$, $Q_{i,t,e}$, and $q_{i,t}$ are the outflow, power generation flow, and waste flow, respectively, of hydropower station $i$ in time period $t$.

f. Power generation flow rate constraint

$$Q_{i,t,e,\min} \le Q_{i,t,e} \le Q_{i,t,e,\max} \tag{8}$$

In the formula, $i$ is the number of the hydropower stations, $i = 1, 2, \ldots, N$, $N$ is the total number of cascade hydropower stations; $t$ is the sequence number of the calculation period, $t = 1, 2, \ldots, T$; $T$ is the total number of time periods, $T = 96$; $Q_{i,t,e,min}$ and $Q_{i,t,e,max}$ are the upper and lower limits of the power generation flow rate of hydropower station $i$ in time period $t$, respectively; $Q_{i,t,e}$ are the power generation flow of hydropower station $i$ in time period $t$.

g. Discharge flow rate constraint

$$Q_{i,t,\min} \le Q_{i,t} \le Q_{i,t,\max} \tag{9}$$

In the formula, $i$ is the number of the hydropower stations, $i = 1, 2, \ldots, N$, $N$ is the total number of cascade hydropower stations; $t$ is the sequence number of the calculation period, $t = 1, 2, \ldots, T$; $T$ is the total number of time periods, $T = 96$; $Q_{i,t,min}$ and $Q_{i,t,max}$ are the upper and lower limits of the discharge flow rate of hydropower station $i$ in time period $t$, respectively.

h. Water level constraint

$$Z_{i,t,\min} \le Z_{i,t} \le Z_{i,t,\max} \tag{10}$$

In the formula, $i$ is the number of the hydropower stations, $i = 1, 2, \ldots, N$, $N$ is the total number of cascade hydropower stations; $t$ is the sequence number of the calculation period, $t = 1, 2, \ldots, T$; $T$ is the total number of time periods, $T = 96$; $Z_{i,t}$ is the water level of hydropower station $i$ in time period $t$; $Z_{i,t,min}$ and $Z_{i,t,\max}$ are the upper and lower limits of the water level of hydropower station $i$ in time period $t$, respectively.

i. Outbound flow rate variation constraint

$$\Delta Q_{i,t} \le \Delta Q_{i,t,\max}, \Delta Q_{i,t} = \left| Q_{i,t} - Q_{i,t-1} \right| \tag{11}$$

In the formula, $i$ is the number of the hydropower stations, $i = 1, 2, \ldots, N$, $N$ is the total number of cascade hydropower stations; $t$ is the sequence number of the calculation period, $t = 1, 2, \ldots, T$; $T$ is the total number of time periods, $T = 96$; $\Delta Q_{i,t}$ and $\Delta Q_{i,t,max}$, respectively, are the variation and maximum variation in the outflow of the power station between time period $t$ and $t - 1$; $Q_{i,t}$ and $Q_{i,t-1}$ are the outflow of the power station in time period $t$ and $t - 1$.

### 2.3.3. Search Strategy

The method proposed in this paper utilizes the sluice operation strategy table, which eliminates a large number of solutions that do not meet operational requirements, shortens the decision space, and avoids the complexities of multi-objective solving by prioritizing the two optimization objectives. Therefore, the use of the traversal search algorithm in this model can meet the requirements of calculation accuracy and efficiency and can avoid getting stuck in local optimal solutions. The specific calculation steps are as follows:

Step 1: Taking the opening of each sluice of each hydropower station at the initial calculation time of the model as the initial solution and deducing the water level and flow rate based on this initial solution.

Step 2: Adjusting the sluice opening one hour in advance if the power station's water level/flow rate is projected to exceed the limit and making an immediate adjustment if less than one hour remains until the moment the limit will be exceeded. If it is necessary to adjust the sluice, recording the time of adjustment.

Step 3: Searching for the set of sluice opening combinations $P_1$ among the sluice operation strategy table based on the water level of the hydropower station at the time of adjustment.

Step 4: Searching for a set of feasible sluice opening combinations $P_2$ that satisfy the various constraints among the set of sluice combinations $P_1$.

Step 5: Calculating the objective function (Equations (1) and (2)) value of each scheme in the set of feasible sluice opening combinations $P_2$, and selecting the best scheme from $P_2$. A flowchart illustrating the process of "online search" is presented in Figure 2.

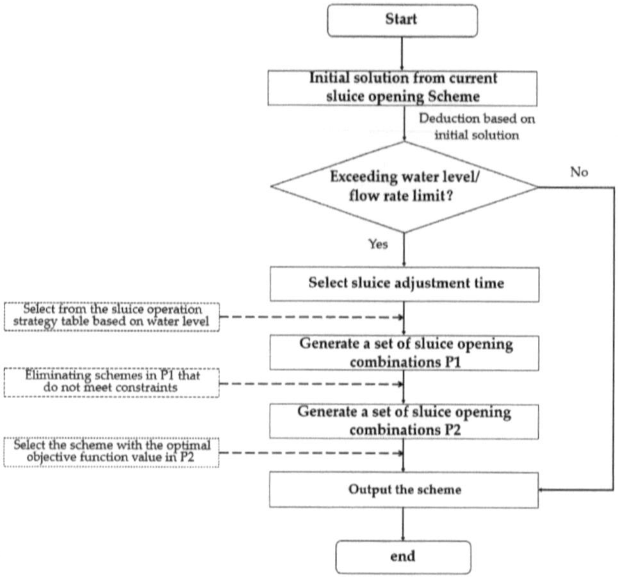


**Figure 2.** Flowchart of "online search" -based sluice operation strategy generation.

## 3. Case Study

### *3.1. Overview of the Study Area*

The Dadu River basin [21] is the fifth largest hydropower base in China. It is the main peak shaving and frequency regulation power source for the Sichuan power grid and a key barrier for flood control in the Yangtze River Basin and surrounding area. The Pubugou, Shenxigou, and Zhentouba power stations (hereinafter referred to as "Pu–Shen–Zhen") in the middle reaches of the Dadu River are mainly used for power generation. On average, they receive grid load adjustment instructions approximately twice per minute. Owing to the complex inflows between areas and cross-sectional tidal restrictions, the three power stations have been afflicted by a long-term power and water mismatch. In particular, the two downstream power stations that were being regulated daily had poor storage performance and short hydropower transmission times, resulting in high-intensity peak shaving and frequency regulation, which produced severe fluctuations in the water level, requiring the frequent use of sluices for water level safety control.

The Pubugou power station flood discharge facilities include three spillways and a flood discharge tunnel. The spillways have three inlet sluices with openings of 12 m × 17 m (width × height). When open, the sluices should be uniformly open, and the frequency of use should be reduced when the water level is below 840 m. The size of the flood discharge tunnel opening is 11.0 m × 11.5 m (width × height), and long-term use with a small degree of opening should be avoided. The Shenxigou flood discharge facilities include three sluices and two flood discharge tunnels. There are three sluices with openings of 7.0 m × 17.0 m (width × height), which adhere to the principle of "uniform symmetry". These sluices have two gates with an outlet size of 9 m × 11.5 m (width × height). The reservoir level should be higher than 650 m when opening, and operations should be avoided in the sluice vibration area. The Zhentouba flood discharge facilities include five sluices, with a total of five openings, each 8.0 m × 16 m (width × height). When operated under low flow rate conditions, asymmetric opening of sluices and excessive discharge by a single sluice should be avoided. The main feature parameters of the three stations are listed in Table 2. The geographical location of the Dadu River Basin is shown in Figure 3.

**Table 2.** Feature parameters of power stations in the middle reaches of the Dadu River Basin.

| Power Station | Regulating Ability | Normal Water Level/m | Dead Water Level /m | Installed Capacity/MW | Spillway Opening Size/m | Discharge Tunnel Opening Size/m | Sluice Opening Size/m |
|---|---|---|---|---|---|---|---|
| Pubugou | Incomplete annual regulation | 850 | 790 | 3600 | 12 × 17 | 11 × 15 | — |
| Shenxigou | Daily regulation | 660 | 655 | 660 | — | 7 × 17 | 9 × 11.5 |
| Zhentouba | Daily regulation | 624 | 618 | 720 | — | — | 8 × 16 |

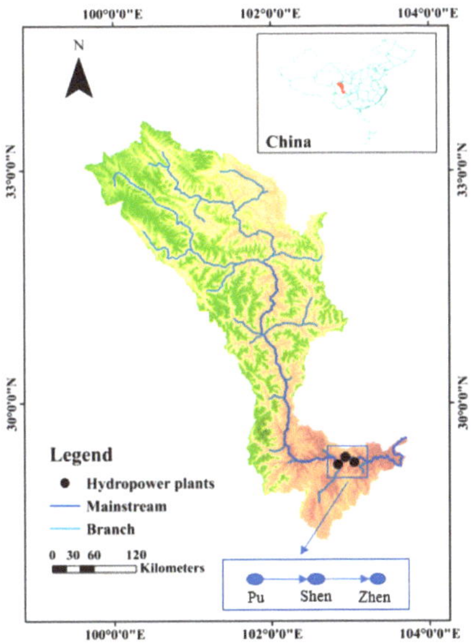


**Figure 3.** Schematic diagram of cascade power stations in the Dadu River Basin.

### 3.2. Data Sources

In this study, the period from 5 July to 5 August 2019 was selected as a typical period. The data used in model calculations include the total load process of the Pu–Shen–Zhen station every 15 min, inflow of Pubugou every 15 min, runoff process between Pubugou and Shenxigou every 15 min, runoff process between Shenxigou and Zhentouba every 15 min, water level process of each station every 15 min, and sluice opening process.

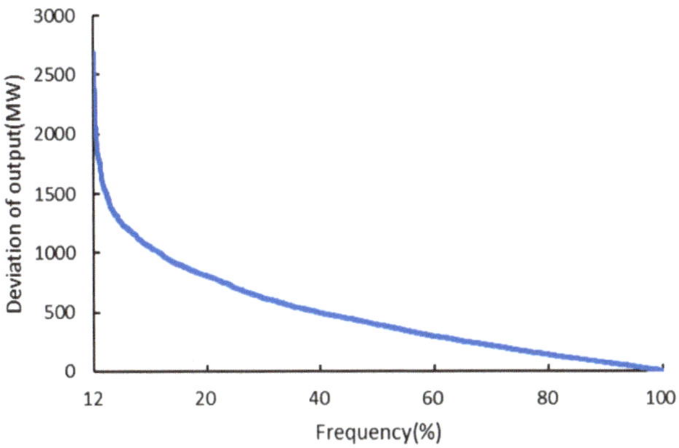


**Figure 5.** Frequency curve of the difference between actual and planned outputs of Pu–Shen–Zhen.

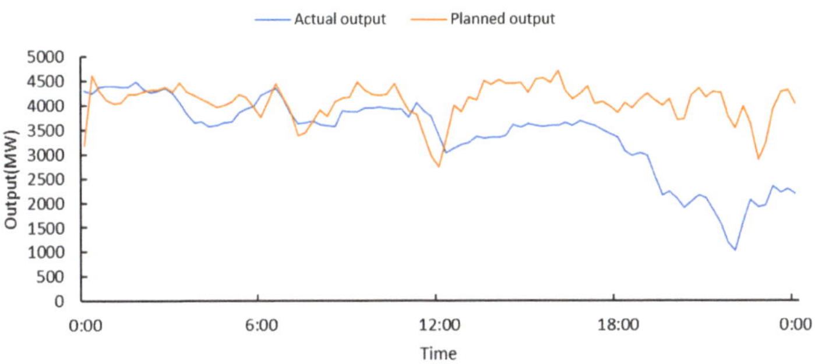


**Figure 6.** Actual and planned outputs of Pu–Shen–Zhen on July 11.

The inherent randomness and unpredictability in the total load of the Pu–Shen–Zhen system pose significant challenges in effectively managing the daily water level of the power station. To accommodate the frequent fluctuations in load instructions, the daily regulation hydropower station experiences sharp rises and falls in its water level, which compromises the station's safety. This unfavorable situation necessitates constant sluice adjustments to regulate the water level. The following section provides a detailed analysis of the water level and sluice scheduling processes.

### 3.3.2. Analysis of Water Level Process

To complete the power generation task of each power station, this study simulated the scheduling process of the power stations for the next 24 h, predicted the risk of each power station exceeding their water level and flow rate limits, and adjusted sluices in advance to meet the regulation and control requirements of each power station. Pubugou has an incomplete annual regulating capacity and requires water level control, with objectives that should be completed as far as possible. Following rolling optimization, the water level of the Pubugou power station was controlled at 840.62 m, which is 0.31 m above the target water level of 840.31 m and meets the water level control requirements. The two stations of Shenxigou and Zhentouba should ensure that the water level is as stable as possible. Before optimization, the water level at Shenxigou fluctuated between 655.11 and 658.85 m, and the lowest water level was only 0.11 m from the dead water level, which posed a significant safety risk. After optimization, the water level at Shenxigou power station was controlled between 656.51 and 658.87 m; thus, the range was reduced to 1.38 m, and the lowest water level was 1.51 m from the dead water level, thereby avoiding the risk of the reservoir emptying. Before optimization, the water level at Zhentouba fluctuated between

618.61 and 622.96 m, and the lowest water level was only 0.61 m from the dead water level. After optimization, the water level of Zhentouba was in the range of 619.15 to 622.95 m, reducing the fluctuation to 0.55 m, and the lowest water level was 1.15 m from the dead water level, which prevented the water level from falling near the dead storage level and improved the scheduling operation of the power station. The process of water level before and after optimization of Pu-Shen-Zhen with 15-min intervals is shown in Figure 7.

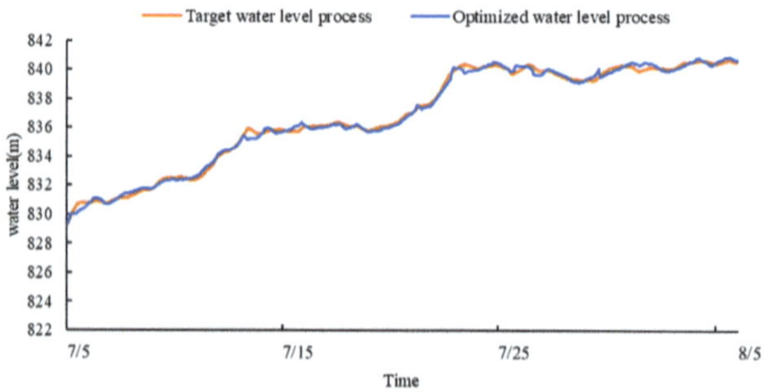


(**a**)

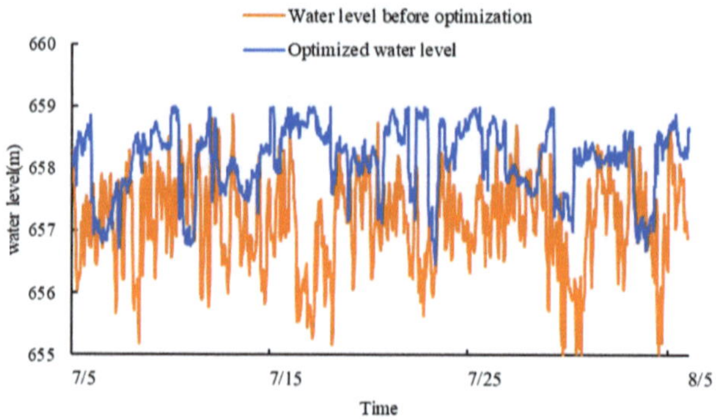


(**b**)

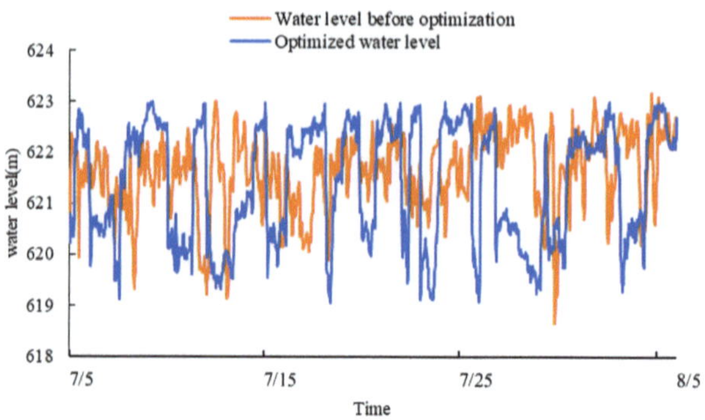


(**c**)

**Figure 7.** (**a**) Water level process of Pubugou power station; (**b**) Water level process of Shenxigou power station; (**c**) Water level process of Zhentouba power station.

### 3.3.3. Analysis of the Sluice Operation Process

The sluice scheduling of power stations is based on traditional manual experience, and the sluice scheduling process often encounters the problems of frequent opening and closing sluices, multiple adjustments, and unsound sluice operation processes. The Shenxigou and Zhentouba stations have small storage capacities, and their reservoir water levels tend to rise and fall steeply due to the influence of the upstream reservoir, resulting in the forced use of sluices to make adjustments, which is not conducive to the safe operation of reservoirs and increases the pressure on workers.

In this study, the combination of offline calculations and online searches not only eliminated irrational feasible sluice solutions and improved the scientific basis of sluice decision-making but also reduced the number of sluice operations and relieved pressure on workers. Following optimization, the total number of sluice movements at the Pu–Shen–Zhen stations was reduced by 43.5%, from 1195 to 675. Sluice movements were reduced by 37.4% at Pubugou, 59.7% at Shenxigou, and 18.1% at Zhentouba. The numbers of sluice adjustments at each power station are listed in Tables 3–5.

**Table 3.** The adjustment times of the sluices at Pubugou hydropower station.

| Mode | Spillway 1# | Spillway 2# | Spillway 3# | Flood Discharging Tunnel | Total |
|---|---|---|---|---|---|
| Before optimization | 4 | 8 | 7 | 144 | 163 |
| After optimization | 11 | 10 | 11 | 70 | 102 |

**Table 4.** The adjustment times of the sluices at Shenxigou hydropower station.

| Mode | Sluice 1# | Sluice 2# | Sluice 3# | Flood Discharging Tunnel 1# | Flood Discharging Tunnel 2# | Total |
|---|---|---|---|---|---|---|
| Before optimization | 132 | 132 | 132 | 110 | 149 | 655 |
| After optimization | 0 | 0 | 0 | 132 | 132 | 264 |

**Table 5.** The adjustment times of the sluices at Zhoutouba hydropower station.

| Mode | Sluice 1# | Sluice 2# | Sluice 3# | Sluice 4# | Sluice 5# | Total |
|---|---|---|---|---|---|---|
| Before optimization | 55 | 0 | 98 | 134 | 90 | 377 |
| After optimization | 65 | 50 | 72 | 50 | 72 | 309 |

In order to fully compare the differences before and after optimization on a long-term scale, this study conducted an optimization of the sluice operation process for one month at 15-min intervals. The specific operation of each sluice with 15-min intervals is shown in Figure 8. In Figure 8, the dashed line represents before optimization, and the solid line represents after optimization. It can be seen from the figure that, before optimization, unsound practices occurred, such as frequent opening and closing within a short period of time and frequent small changes in the extent to which sluices were opened. At Shenxigou power station, in particular, the degree of openness of all five sluices repeatedly changed. After optimization, the operations of sluices at each power station became more regular, and the opening of each sluice became more stable, enhancing the scientificity and accuracy of sluice decision-making.

The current gate scheduling process predominantly relies on manual experience-based scheduling. However, due to the inherent limitations of manual scheduling, dispatchers face challenges in accurately predicting the flow and water level trends of the future cascade power station group. Consequently, they struggle to select the most optimal gate operation plan, leading to the undesirable phenomenon of repeated opening adjustments and frequent gate openings and closures within the power station.

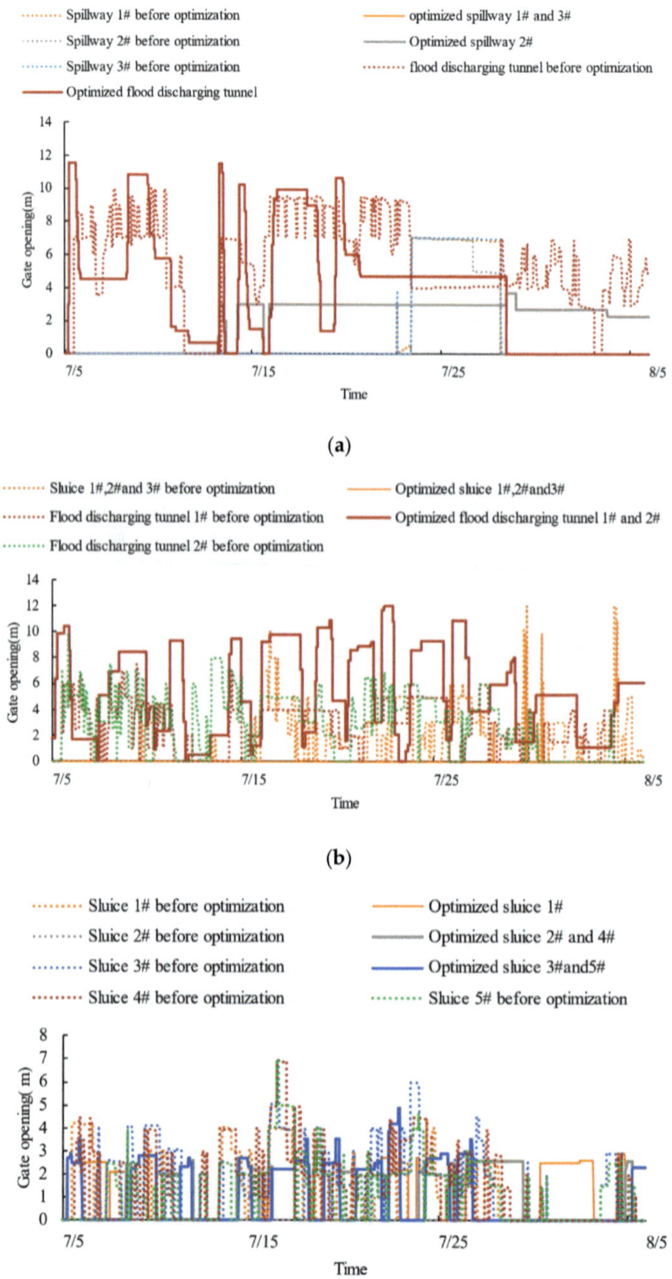


(c)

**Figure 8.** (**a**) Pubugou power station sluice operation process; (**b**) Shenxigou power station sluice operation process; (**c**) Zhentouba power station sluice operation process.

To provide a comprehensive analysis, we consider 8 July 2019 as a representative day, focusing on the intraday gate scheduling process. At the Shenxigou hydropower station, at

01:00, the gate openings for the 1# and 2# spillway tunnels were initially increased from 5.5 m to 6.5 m and 7.5 m, respectively. However, after maintaining this configuration for 30 min, the openings were subsequently decreased. Furthermore, at 10:00, the opening of the 2# flood discharge tunnel was adjusted from 5.5 m to closed. However, after a brief 15-min interval, the gate was reopened to 3.0 m. Such inconsistencies and frequent adjustments in gate openings highlight the limitations of manual scheduling, which fails to consider future water flow trends and optimal gate configurations.

Alternatively, by leveraging a model that comprehensively considers future water flow trends and various gate combinations, a more refined approach to gate scheduling can be achieved. The model's optimized gate opening adjustments only occur once during a 24-h period, specifically at 11:30, effectively avoiding the irrational occurrence of frequent gate opening changes during actual dispatching. Similarly, at the Zhentouba hydropower station, the impact of upstream gate operations in the actual dispatching process resulted in frequent changes in gate openings at 01:30, 11:00, and 20:00. The specific process of sluice operation with 15-min intervals on 8 July 2019 is shown in Figure 9, and the comparison parts mentioned above have been marked with red frame in the Figure 9. However, through the application of the model optimization, the number of gate opening adjustments is significantly reduced, resulting in a more regular and rational gate operation method.

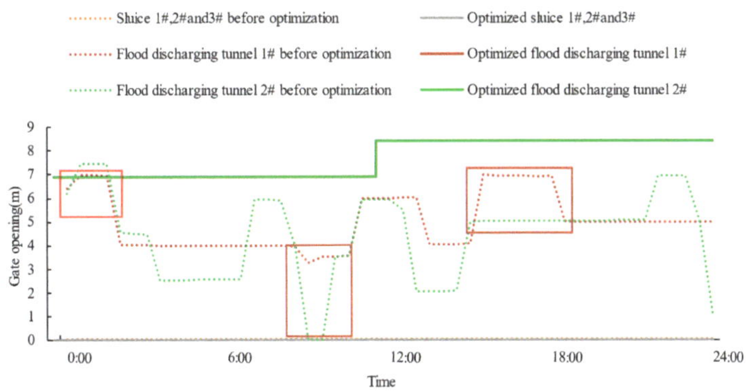


(a)

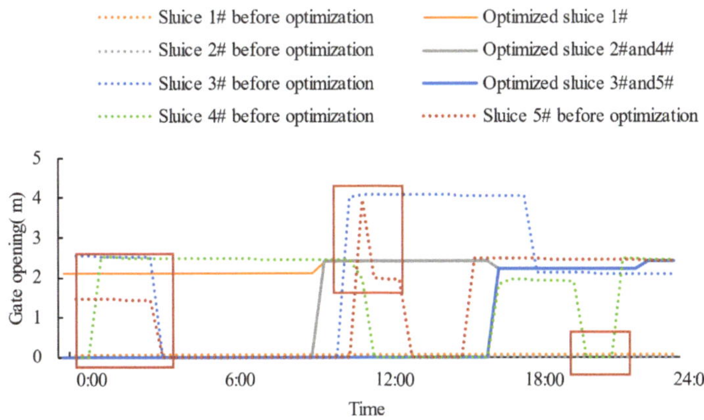


(b)

**Figure 9.** (**a**) Shenxigou power station sluice operation process on July 8 and (**b**) Zhentouba power station sluice operation process on July 8.

By addressing the shortcomings of manual scheduling and embracing the advancements offered by model optimization, the gate scheduling process can achieve greater accuracy, efficiency, and stability, minimizing unnecessary fluctuations and improving the overall operational performance of the power station.

## 4. Conclusions

In this study, in view of the frequent operation of sluices in the group of cascade power stations under high-intensity peak shaving and frequency regulation, a joint-optimized operation model of sluices in the group was established. A sluice operation strategy table was created via offline calculations to ensure rational and timely decision-making with respect to sluice opening. Based on the initial state of each power station and water forecast information, we simulated the water level of each reservoir for the next 24 h and then conducted an online search for the optimal sluice operation strategy to automate sluice opening at each station. The main conclusions are summarized as follows:

(1) The method proposed in this paper efficiently addressed the problem of frequent sluice operations at the Pu–Shen–Zhen cascade power stations under high-intensity peak shaving and frequency regulation and enhanced the scientific basis of decision-making of the power station group scheduling. The total number of times the sluices of the cascade power stations were used was reduced from 1195 to 675, a reduction of 43.5%.

(2) The method proposed in this paper can reduce the fluctuation amplitude of water level in daily regulating power stations and ensure the safety of cascade power station operation while meeting the requirements of water level control in the leading power station. The optimized water level of Pubugou differs from the target water level by a mere 0.31 m, whereas the water level fluctuations of the two downstream daily regulating power stations of Shenxigou and Zhentouba were reduced by 1.38 m and 0.55 m, respectively.

In recent years, the large-scale development and connection of renewable energy has placed greater requirements on cascade hydropower scheduling. In the future, the method proposed in this paper can be improved through the following aspects:

(1) Advanced algorithms such as series methods [22] and heuristic algorithms can be used to solve the model proposed in this paper. The use of data mining [23–25], artificial intelligence [26–28], and digital twins [29,30] could also be further explored in the future to further improve the accuracy and timeliness of scheduling decisions.

(2) This study was conducted in the Dadu River Basin of China, and can also be conducted in multiple basins in the future to further improve the scientific and universal nature of this method.

**Author Contributions:** Conceptualization, S.M. and X.W.; Methodology, S.M.; Software, Y.L.; Validation, T.Q. and J.L.; Formal analysis, X.W.; Investigation, T.Q., J.L.; resources, S.M., T.Q. and J.L.; data curation, Y.L.; writing—original draft preparation, S.M.; writing—review and editing, X.W.; visualization, S.M.; supervision, X.W.; project administration, S.M. and X.W.; funding acquisition, X.W. All authors have read and agreed to the published version of the manuscript.

**Funding:** This research was funded by the National Key Research and Development Program of China, grant number 2019YFE0105200, Intelligent Scheduling and Optimal Operation of Cascade Hydropower Stations based on Spatiotemporal Big Data, 2020–2022.

**Data Availability Statement:** All data and models that support the findings of this study are available from the corresponding author upon reasonable request.

**Acknowledgments:** We have immense gratitude for the people of the Dadu River Basin Production Command Center for their funding and data support. We are very grateful for the technical support provided by the College of Water Conservancy and Hydropower Engineering at Hohai University.

**Conflicts of Interest:** Author Shiyu Mou, Tian Qu, Jia Li was employed by the company Dadu River Basin Production Command Center, China Energy Investment. The remaining authors declare that the research was conducted in the absence of any commercial or financial relationships that could be construed as a potential conflict of interest.

## References


1. Galletti, A.; Avesani, D.; Bellin, A.; Majone, B. Detailed simulation of storage hydropower systems in large alpine watersheds. *J. Hydrol.* **2021**, *603*, 127125. [CrossRef]
2. Nazemi, A.; Wheater, H.S. On inclusion of water resource management in earth system models—Part 1: Problem definition and representation of water demand. *Hydrol. Earth Syst. Sci.* **2015**, *19*, 33–61. [CrossRef]
3. He, S.; Gao, H.; Liu, J.; Chen, Z. Distribution system planning considering peak shaving of energy stations. *Appl. Energy* **2022**, *312*, 118692. [CrossRef]
4. Wang, P.; Yuan, W.; Su, C.; Wu, Y.; Lu, L.; Yan, D.; Wu, Z. Short-term optimal scheduling of cascade hydropower plants shaving peak load for multiple power grids. *Renew. Energy* **2022**, *184*, 68–79. [CrossRef]
5. Rana, M.M.; Atef, M.; Sarkar, M.R.; Uddin, M.; Shafiullah, G.M. A Review on Peak Load Shaving in Microgrid-Potential Benefits, Challenges, and Future Trend. *Energies* **2022**, *15*, 2278. [CrossRef]
6. De Ladurantaye, D.; Gendreau, M.; Potvin, J.Y. Optimizing profits from hydroelectricity production. *Comput. Oper. Res.* **2009**, *36*, 499–529. [CrossRef]
7. Avesani, D.; Zanfei, A.; Marco, N.D.; Galletti, A.; Ravazzolo, F.; Righetti, M.; Majone, B. Short-term hydropower optimization driven by innovative time-adapting econometric model. *Appl. Energy* **2022**, *310*, 118510. [CrossRef]
8. Chen, S.; Zhang, Y.; Dan, L.I. Study on constant outflow model for reservoir flood control operation and its application. *Adv. Water Sci.* **2021**, *32*, 683–693. [CrossRef]
9. Abdelal, Q.; Al-Rawabdeh, A.; Al Qudah, K.; Hamarneh, C.; Abu-Jaber, N. Hydrological assessment and management implications for the ancient Nabataean flood control system in Petra, Jordan. *J. Hydrol.* **2021**, *601*, 126583. [CrossRef]
10. Jain, S.K.; Shilpa, L.S.; Rani, D.; Sudheer, K.P. State-of-the-art review: Operation of multi-purpose reservoirs during flood season. *J. Hydrol.* **2023**, *618*, 129165. [CrossRef]
11. Wang, W.; Tian, W.; Chau, K.; Zang, H.; Ma, M.; Feng, Z.; Xu, D. Multi-Reservoir Flood Control Operation Using Improved Bald Eagle Search Algorithm with ε Constraint Method. *Water* **2023**, *15*, 692. [CrossRef]
12. He, J.; Liu, S.-M.; Chen, H.-T.; Wang, S.-L.; Guo, X.-Q.; Wan, Y.-R. Flood Control Optimization of Reservoir Group Based on Improved Sparrow Algorithm (ISSA). *Water* **2023**, *15*, 132. [CrossRef]
13. Wang, Q.; Li, A.; Lu, C. Multi-objective Optimization of Large-scale Multi-reservoir Flood Control Operation. *J. Phys. Conf. Ser.* **2022**, *2333*, 012016. [CrossRef]
14. Xu, Z.X.; Mo, L.; Zhou, J.Z.; Zhang, X. Optimal dispatching rules of hydropower reservoir in flood season considering flood resources utilization. A case study of Three Gorges Reservoir in China. *J. Clean. Prod.* **2023**, *388*, 135975. [CrossRef]
15. Huang, X.; Xu, B.; Zhong, P.A.; Yao, H.; Yue, H.; Zhu, F.; Lu, Q.; Sun, Y.; Mo, R.; Li, Z. Robust multi-objective reservoir operation and risk decision-making model for real-time flood control coping with forecast uncertainty. *J. Hydrol.* **2022**, *605*, 127334. [CrossRef]
16. Zhu, F.L.; Zhong, P.A.; Sun, Y.M.; Yeh, W.G. Real-Time Optimal Flood Control Decision Making and Risk Propagation Under Multiple Uncertainties. *Water Resour. Res.* **2018**, *53*, 10635–10654. [CrossRef]
17. Su, C.; Wang, P.; Yuan, W.; Cheng, C.; Zhang, T.; Yan, D.; Wu, Z. An MILP based optimization model for reservoir flood control operation considering spillway sluice scheduling. *J. Hydrol.* **2022**, *613*, 128483. [CrossRef]
18. Kim, Y.G.; Sun, B.Q.; Kim, P.; Jo, M.B.; Pak, G.H. A study on optimal operation of sluice-controlled reservoir system for flood control based on PSO algorithm combined with rearrangement method of partial solution groups. *J. Hydrol.* **2021**, *593*, 125783. [CrossRef]
19. Zhiqiang, J.; Hui, Q.; Changming, J.; Zhongkai, F.; Jianzhong, Z. Two Dimension Reduction Methods for Multi-Dimensional Dynamic Programming and Its Application in Cascade Reservoirs Operation Optimization. *Water* **2017**, *9*, 634. [CrossRef]
20. Song, J.H.; Her, Y.; Kang, M.S. Estimating Reservoir Inflow and Outflow from Water Level Observations Using Expert Knowledge: Dealing with an Ill-Posed Water Balance Equation in Reservoir Management. *Water Resour. Res.* **2022**, *58*, e2020WR028183. [CrossRef]
21. Sun, Y.; Zhu, F.; Chen, J.; Li, J. Risk Analysis for Reservoir Real-Time Optimal Operation Using the Scenario Tree-Based Stochastic Optimization Method. *Water* **2018**, *10*, 606. [CrossRef]
22. Turkyilmazoglu, M. Nonlinear problems via a convergence accelerated decomposition method of adomian. *Comput. Model. Eng. Sci.* **2021**, *127*, 1–22. [CrossRef]
23. Huang, K.; Yuan, J.; Zhou, Z.; Zheng, X. Analysis and evaluation of heat source data of large-scale heating system based on descriptive data mining techniques. *Energy* **2022**, *251*, 123834. [CrossRef]
24. Ruidas, D.; Saha, A.; Chowdhuri, I.; Pal, S.C.; Towfiqul Islam, A.R.M. Application of novel data-mining technique based nitrate concentration susceptibility prediction approach for coastal aquifers in India. *J. Clean. Prod.* **2022**, *346*, 131205. [CrossRef]
25. Attari, M.; Ejlaly, B.; Heidarpour, H.; Ala, A. Application of Data Mining Techniques for the Investigation of Factors Affecting Transportation Enterprises. *IEEE Trans. Intell. Transp. Syst.* **2022**, *23*, 9184–9199. [CrossRef]
26. Feng, Z.K.; Shi, P.F.; Yang, T.; Niu, W.J.; Zhou, J.Z.; Cheng, C.T. Parallel cooperative search algorithm and artificial intelligence method for streamflow time series forecasting. *J. Hydrol.* **2022**, *606*, 127434. [CrossRef]
27. Cui, F.; Al-Sudani, Z.A.; Hassan, G.S.; Afan, H.A.; Ahammed, S.J.; Yaseen, Z.M. Boosted artificial intelligence model using improved alpha-guided grey wolf optimizer for groundwater level prediction. Comparative study and insight for federated learning technology. *J. Hydrol.* **2022**, *606*, 127384. [CrossRef]

28. Hadadi, F.; Moazenzadeh, R.; Mohammadi, B. Estimation of actual evapotranspiration. A novel hybrid method based on remote sensing and artificial intelligence. *J. Hydrol.* **2022**, *609*, 127774. [CrossRef]
29. Zhao, D.; He, Q.; Yu, J.; Guo, M.; Fu, J.; Li, X.; Ni, M. A data-driven digital-twin model and control of high temperature proton exchange membrane electrolyzer cells. *Int. J. Hydrogen Energy* **2022**, *47*, 8687–8699. [CrossRef]
30. Huang, Z.F.; Soh, K.Y.; Islam, M.R.; Chua, K.J. Digital twin driven life-cycle operation optimization for combined cooling heating and power-cold energy recovery (CCHP-CER) system. *Appl. Energy* **2022**, *324*, 119774. [CrossRef]

*water*

MDPI

*Article*

# Development and Deployment of a Virtual Water Gauge System Utilizing the ResNet-50 Convolutional Neural Network for Real-Time River Water Level Monitoring: A Case Study of the Keelung River in Taiwan

**Jui-Fa Chen *, Yu-Ting Liao and Po-Chun Wang**

Department of Computer Science and Information Engineering, Tamkang University, New Taipei City 251301, Taiwan; qwe123w321@gmail.com (Y.-T.L.); davidvw1298@gmail.com (P.-C.W.)

* Correspondence: alpha@mail.tku.edu.tw; Tel.: +886-922-557-739

**Abstract:** Climate change has exacerbated severe rainfall events, leading to rapid and unpredictable fluctuations in river water levels. This environment necessitates the development of real-time, automated systems for water level detection. Due to degradation, traditional methods relying on physical river gauges are becoming progressively unreliable. This paper presents an innovative methodology that leverages ResNet-50, a Convolutional Neural Network (CNN) model, to identify distinct water level features in Closed-Circuit Television (CCTV) river imagery of the Chengmei Bridge on the Keelung River in Neihu District, Taiwan, under various weather conditions. This methodology creates a virtual water gauge system for the precise and timely detection of water levels, thereby eliminating the need for dependable physical gauges. Our study utilized image data from 1 March 2022 to 28 February 2023. This river, crucial to the ecosystems and economies of numerous cities, could instigate a range of consequences due to rapid increases in water levels. The proposed system integrates grid-based methods with infrastructure like CCTV cameras and Raspberry Pi devices for data processing. This integration facilitates real-time water level monitoring, even without physical gauges, thus reducing deployment costs. Preliminary results indicate an accuracy range of 83.6% to 96%, with clear days providing the highest accuracy and heavy rainfall the lowest. Future work will refine the model to boost accuracy during rainy conditions. This research introduces a promising real-time river water level monitoring solution, significantly contributing to flood control and disaster management strategies.

**Keywords:** ResNet-50; Convolutional Neural Network; water level detection; river monitoring system; real-time monitoring system; virtual water gauge; grid-based

## 1. Introduction

The escalating impact of climate change, marked by a worldwide increase in severe weather events, particularly unpredictable and rapid fluctuations in river water levels, necessitates the development of reliable, real-time water level detection systems [1–3]. In Taiwan, a country experiencing a heightened frequency of typhoons and heavy rainfall [4,5], the increased risk of rising river water levels underscores the need for remote hydrological monitoring, especially during typhoons or significant precipitation [6]. This situation highlights the vital importance of automated water level measurement systems [7–10].

Traditional methods, which heavily rely on physical river gauges, are becoming increasingly unreliable due to environmental degradation [11–15]. An alternative approach using CCTV cameras to monitor water gauges installed in significant rivers and flood-prone areas has been explored. Yu [16] proposed a differencing image technique that detects minor changes in water levels by analyzing the Region of Interest (ROI) between previous and current frames and applying the Otsu threshold method. However, the robustness of this method under different illuminations and locations still needs to be tested. Kim [17] developed a cloud-based system, the River Eye Image Water Level Gauge, which integrates video surveillance for river flow and water level measurements. This system is currently undergoing testing at four sites. Hiroi [18] presented a water-level sensor system that uses infrared image processing for real-time river-level monitoring and accurate flood prediction in urban areas. Pan [19] developed a low-cost unmanned surveillance system that uses a map-based web service, video cameras, water level analyzers, and wireless communication routers for real-time water level measurements. The deep learning-based method demonstrated superior performance in terms of accuracy and stability. Sabbatini [20] proposed a computer vision solution for automatic river water-level monitoring, showing excellent performance in discerning frame quality, especially during nighttime. Narayanan [21] introduced a method that uses participatory sensing and computer vision to estimate flood levels.

However, the absence of gauges in some rivers and inaccurate readings due to inadequate maintenance hinder precise water level detection [22,23]. As illustrated in Figure 1, poorly maintained gauges often present unclear numerical readings, preventing image recognition technology from accurately determining the current water level and leading to potential misjudgments. This highlights an urgent need for further research to mitigate these issues.

**Figure 1.** It is challenging to recognize unclear numerical readings in river gauge images. Source: https://fmg.wra.gov.tw/fmgp/ccd_proxy?sn=55 (accessed on 17 October 2022).

To address this challenge, this paper introduces an innovative approach that utilizes existing CCTV footage to establish a virtual water gauge. This method subsequently applies image processing techniques to determine the current water level of the river. The proposed model, which employs ResNet-50 [24–26], a Convolutional Neural Network (CNN) model [27–31], is trained using data derived from the CCTV river imagery of the Chengmei Bridge on the Keelung River in Neihu District, Taiwan. This data was collected from 1 March 2022 to 28 February 2023.

The Keelung River, an essential water system in Northern Taiwan, has an approximate length of 96 km and a catchment area of around 512 square kilometers. It traverses prominent cities in Northern Taiwan, including Taipei, New Taipei, and Keelung, significantly influencing these regions' geographical and economic landscapes.

The Keelung River has been the site of numerous severe floods throughout the years, marked by notably devastating events brought on by Typhoon Lynn in October 1987, Typhoon Winnie in August 1997, Typhoon Xangsane in October 2000, and Typhoon Nari in September 2001. These typhoons unleashed torrential rains, leading to extensive flooding within the Keelung River basin. For example, Typhoon Xangsane in 2000 resulted in an inundation of approximately 465 hectares spanning various districts in Taipei City, New

Taipei City, and Keelung City, causing 59 fatalities and flooding around 10,000 households. Moreover, Typhoon Nari in 2001 led to severe flooding in the Keelung River basin, submerging numerous areas in Taipei City, New Taipei City, and Keelung City, resulting in 104 deaths and approximately 20,000 flooded households.

The most recent incident occurred on 16 October 2022, when Typhoon Nisha swept across the Keelung River basin. This event led to a dramatic surge in the river's water level, peaking at 5.04 m, exceeding the alert level by two meters. This occurrence highlighted the Keelung River basin's vulnerability to flood risks under extreme weather conditions. Such disasters can significantly impact residents in low-lying areas and the surrounding communities, resulting in property damage, road traffic disruptions, and casualties. Therefore, implementing effective flood prevention measures and disaster management strategies in the Keelung River basin is critically important to mitigate the impacts of future extreme weather events.

This study is particularly significant for the Keelung River, especially near the Chengmei Bridge. Its geographical location, impact on the surrounding cities, and environmental challenges make it an ideal site for testing our innovative water level detection methods. Influenced by heavy rainfall, the river's water level fluctuations allow us to refine and test our system to enhance the region's flood prevention and disaster management strategies.

A unique feature of this study is the utilization of the existing CCTV infrastructure installed across numerous rivers. By integrating a cost-effective hardware device, Raspberry Pi [32–34], the pretrained grid-based virtual water gauge model can be executed to determine the current river water level. This approach significantly reduces the cost of establishing river water level monitoring facilities and can be readily implemented in various locations.

The paper is structured as follows: The subsequent section will detail the methodology, including the model's training and operation. The following section will present the study's results, including its effectiveness during heavy rainfall. The concluding section will discuss the implications of the study, its limitations, and future research directions. This innovative approach represents a significant advancement in flood control and disaster management, offering considerable potential for enhancing cities' resilience to flooding and other water-related disasters.

## 2. Materials and Methods

In this section, we elaborate on developing and implementing three primary models designed to enhance the accuracy and reliability of river water level monitoring. Firstly, the Grid Selection Model uses CCTV footage and image processing techniques to identify the optimal grid for establishing a virtual water gauge. Secondly, the Grid State Recognition Model accurately determines the state of the selected grid, categorizing it as devoid of water, partially filled with water, or filled with water. This categorization is crucial for determining the river's current water level. Lastly, the Water Level Calculation Model calculates the water level height of the virtual water gauge, converting the grid's state into a numerical value that represents the river's current water level. These models work collectively to provide accurate, real-time water level detection. Their methodologies, execution, and collaborative functioning are detailed in the following sections and depicted in Figure 2.

### 2.1. Grid Selection Model

The Grid Selection Model, a critical system component, is designed to select representative grids from CCTV footage. These grids effectively reflect water level variations, forming the foundation for a virtual water gauge. The model operates in five stages. River imagery and weather data are initially collected to guide the subsequent processes. Following this, the recognition area within the imagery is identified, focusing on relevant areas to enhance efficiency and precision. The images are then divided into smaller, manageable grids for individual examination. These images undergo dynamic binarization, transforming into a binary format for a streamlined analysis. The final stage involves selecting grids that accurately reflect water level changes. These grids provide data for precise water level monitoring. Each stage is meticulously fine-tuned to maximize the model's effectiveness, with an in-depth exploration of each stage offering a comprehensive understanding of the Grid Selection Model's operation.

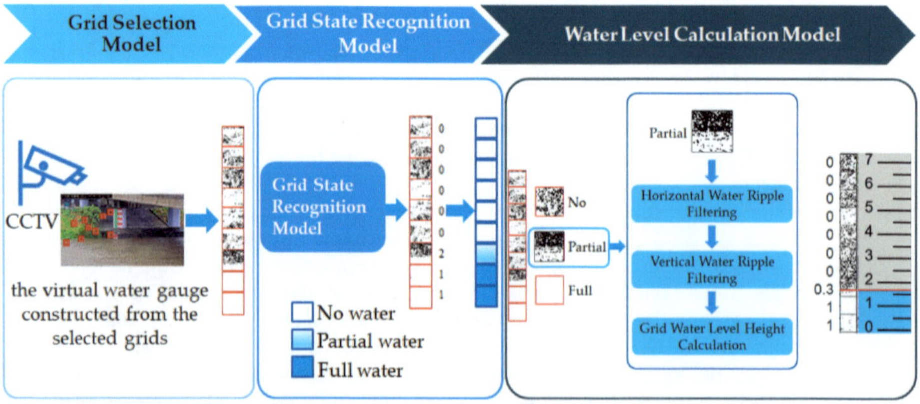


**Figure 2.** Diagrammatic representation of the methodologies and interactions of the three models.

### 2.1.1. Collection of River Imagery and Weather Data

The acquisition of river imagery is a critical phase in our methodology. For a meaningful analysis, gathering images representing various water level changes is essential. Notably, images captured after rainfall events, which cause water levels to surge, are particularly valuable, as they distinctly record fluctuations in water levels.

Simultaneously, we collect weather information specific to the river's location, encompassing the current weather conditions and sunrise and sunset times. This data aid in refining the threshold for binarization during the dynamic image binarization stage, considering the environmental factors in the imagery. These weather data are typically acquired from meteorological observation stations or similar entities.

For this study, river imagery was sourced from the CCTV at the Chengmei Bridge on the Keelung River in Taiwan. This open data was gathered from 1 March 2022 to 28 February 2023 via the Water Resources Agency's Water Situation Image Monitoring Station's cloud service platform, a Ministry of Economic Affairs subsidiary. The CCTV river images, with a resolution of 1920 × 1080, are updated every minute, as illustrated in Figure 3.

Additionally, weather data, including hourly weather conditions and sunrise and sunset timings from 1 March 2022 to 28 February 2023, was compiled for the Neihu District of Taipei City, where the Chengmei Bridge is situated. These data were sourced from the Central Weather Bureau's website, Taiwan's Ministry of Transportation and Communications Division.

This comprehensive collection of river imagery and weather data forms a solid foundation for the subsequent stages of our Grid Selection Model, significantly enhancing the accuracy of our water level monitoring system.

**Figure 3.** CCTV river imagery of the Chengmei Bridge on the Keelung River in Neihu District, Taiwan. Source: https://fmg.wra.gov.tw/fmgp/ccd_proxy?sn=40 (accessed on 12 March 2022).

### 2.1.2. Determination of the Recognition Area

The initial step in establishing a virtual water gauge within the CCTV footage of a river involves selecting an appropriate region. This region is the basis for filtering suitable unit grids to construct the virtual water gauge. The selection of this region is critical and should align with the actual water level fluctuations in the river.

This designated region assists in selecting the ideal grid units that constitute the virtual water gauge. The lower limit of this region corresponds with the lowest point of the actual water level, represented as zero on the water gauge. Conversely, the upper limit of this region aligns with the highest point of the water level. This process and the corresponding water levels are depicted in Figure 4.

**Figure 4.** Outline of the recognition region for the virtual water gauge in the river imagery.

In locations equipped with water gauges, the proposed grid-based virtual water level can be directly inferred from the height of the water gauge, as demonstrated in the image. However, for areas lacking water gauges, it becomes necessary to calculate the conversion formula and parameters that link the virtual water level to the actual water level. This process requires on-site measurements. Using this information, the current water level in the river can be accurately determined, even in locations where water gauges are absent or the readings are obscured.

In this case, when identifying the recognition area from the collected CCTV river imagery at Chengmei Bridge, the lowest point within the recognition area corresponds with the part of the river where the water level is 0 m on the water gauge. Conversely, the highest point aligns with the highest point on the river water gauge, situated at 7.5 m. This is illustrated in Figure 5.

**Figure 5.** Outline of the recognition area in the CCTV river imagery at Chengmei Bridge. Source: https://fmg.wra.gov.tw/fmgp/ccd_proxy?sn=40 (accessed on 14 March 2022).

### 2.1.3. Image Gridification

The next step in creating a virtual water gauge from the river imagery involves segmenting the identified area within the image into a grid format. The dimensions of these grids should closely correspond to the pixel count in the image that represents the unit height of the actual water level. With appropriate segmentation, we can select the most suitable grids to form the virtual water gauge.

For the ensuing grid selection, we employ the ResNet50 model, which is implemented using the TensorFlow package in Python, to identify the features of the grid images. This methodology will assist us in selecting grids suitable for constructing the virtual water gauge. It is imperative to note that the image input size for the ResNet50 model is 224 × 224. Therefore, we must ensure that the grid dimensions do not introduce any distortions or alterations to the inherent features of the grid during the scaling process to match the required input size.

To this end, the number of pixels corresponding to the grid height, denoted as $\text{Grid}_{\text{Pixel}}$, is determined using Equation (1). Assuming the unit height of the actual water level corresponds to P pixels in the image, the formula is as follows:

$$\text{Grid}_{\text{Pixel}} = \left\{7 \times 2^{\text{N}} \,|\, \text{N} = \lfloor \log_2^{\text{P}/7} \rfloor \right\} \tag{1}$$

N is an exponential term in this equation that adjusts the grid size to closely approximate P. This formula aims to align the grid height ($\text{Grid}_{\text{Pixel}}$) with the image height (P) corresponding to the unit height of the actual water level. Simultaneously, it ensures that, when increased by a power of 2 (to the Nth power), the final size is 224 × 224. The choice of 7 as the base in this formula is informed by the fact that, when 224 is continuously halved, the smallest value achievable is 7.

In this study, involving the river imagery from Chengmei Bridge, an actual 1 m corresponds to 65 pixels in the image. According to Equation (1), where P is given as 65, the value of $\text{Grid}_{\text{Pixel}}$ can be calculated by using Equation (1): $\text{Grid}_{\text{Pixel}} = \{7 \times 2^{\text{N}} \,|\, \text{N} = \lfloor \log_2^{65/7} \rfloor = 3\} = 7 \times 2^3 = 56$, which simplifies as the $\text{Grid}_{\text{Pixel}}$ is 56.

Therefore, when performing image gridification, a grid height unit of 56 pixels should be used for processing. Figure 6 shows the recognition region partitioned into grids of uniform size, each approximating the unit height of the real-world water level.

### 2.1.4. Dynamic Image Binarization

The binarization process is crucial in image recognition, particularly when recognizing river water levels [35,36]. However, employing dynamic image binarization is necessary due to several potential complications. Traditional static thresholding methods often need to perform better when applied to images captured under various environmental conditions. A single threshold value may not be universally suitable for different lighting and weather conditions or times of the day. This can lead to suboptimal binarization and, consequently, decreased recognition accuracy. Additionally, a static approach needs more

## 3.3. Analysis and Discussion

### 3.3.1. Load Process Analysis

In conducting an exhaustive examination of the actual and planned outputs of the Pu–Shen–Zhen cascade power station over the course of a month, a conspicuous dissimilarity between the two datasets comes to the forefront. The maximum deviation between the actual and planned outputs of the Pu–Shen–Zhen cascade power station amounts to a substantial 2682 MW, which represents a significant deviation of 68.4% from the initially projected output. Further analysis reveals an average deviation of 488 MW, indicative of a significant variance equivalent to 14.6% of the planned output value when considering the average deviation percentage. Remarkably, close to 12% of the analyzed time period exhibits deviations surpassing the threshold of 1000 MW.

Additionally, the actual output process exhibits highly volatile fluctuations, characterized by substantial gaps between peak and valley levels, with deviations exceeding 1500 MW, accounting for a notable 32% of the maximum output. These findings provide compelling evidence of the arduous tasks undertaken by the three stations within the Pu–Shen–Zhen cascade power system, which are responsible for high-intensity peak shaving and frequency regulation activities. Given the elevated frequency and magnitude of load changes, these stations face the formidable challenge of ensuring grid stability and operational efficacy. A comprehensive visualization of these findings is provided in Figures 4 and 5 for reference. Among them, Figure 4 shows the Pu-Shen-Zhen output variation process with 15-min intervals, 7/5 represents 5 July 2019 at 00:00 (This is also the same in the following figures), and Figure 5 shows the distribution frequency of deviation between planned output and actual output.

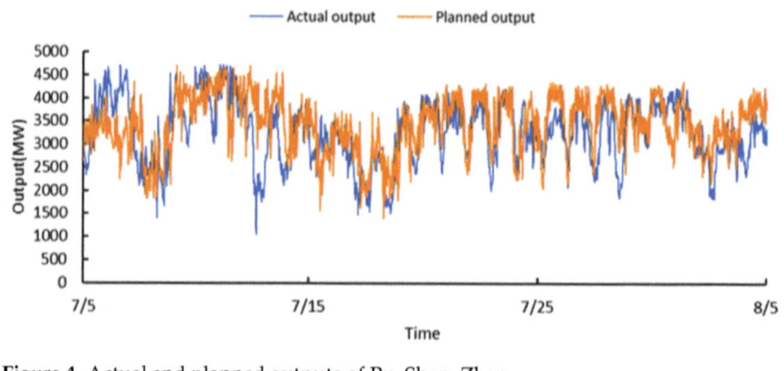


**Figure 4.** Actual and planned outputs of Pu–Shen–Zhen.

The temporal variation between the actual and planned output processes exhibits an irregular and capricious nature. To conduct a meticulous analysis, a representative day, 11 July 2019, was selected for a detailed examination. Notably, throughout this particular day, the planned output showcases a persistent fluctuation in the vicinity of 4000 MW. It is worth highlighting that the planned output demonstrates a commendable convergence with the actual output, harmoniously synchronized until the early afternoon, prior to 12:00. However, subsequent to this temporal threshold, a subtle but discernible decline can be seen in the actual output. Finally, after 18:00, a marked precipitous descent is encountered, in stark contrast to the planned output's swift resumption to its prior trajectory following a minor ephemeral dip. As a consequence, a substantial deviation between the actual and planned outputs emerges, reflecting the intricacies inherent in the dynamic interplay between these two factors. The variation between actual and planned outputs at different times on 11 July 2019, with 15-min intervals, is depicted in Figure 6 (This is also the same in the following figures).

flexibility to adapt in real time to environmental changes, which can further compromise the reliability of the image recognition process.

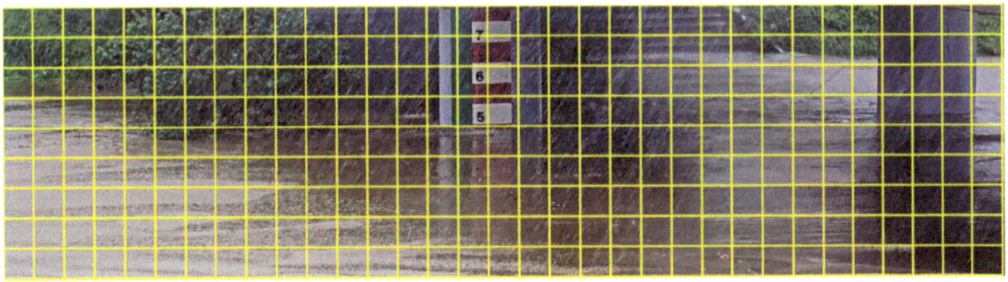

**Figure 6.** Grid partitioning of the river imagery at Chengmei Bridge, size = 56 × 56 pixels. Source: https://fmg.wra.gov.tw/fmgp/ccd_proxy?sn=40 (accessed on 14 March 2022).

To illustrate this, refer to Figure 7, which showcases images taken at 8:00 a.m. at the Nanhu Bridge over the Keelung River under different weather conditions. These images were binarized using a fixed threshold of 150. Figure 7a depicts a sunny day, and Figure 7b represents an overcast day. The results demonstrate that a fixed threshold can discern the water gauge's numbers in Figure 7b. However, in Figure 7a, some numbers need to be more identifiable, indicating the limitations of using a fixed threshold under varying weather conditions.

(a) (b)

**Figure 7.** Binarized images at 8:00 a.m. at Nanhu Bridge on Keelung River using a fixed threshold of 150. (**a**) Binarized image on a sunny day. (**b**) Binarized image on a cloudy day. Source: https://fmg.wra.gov.tw/FMGP/SingeView?sn=15893&ft=C&ht=0 ((**a**) accessed on 10 April 2022, (**b**) accessed on 15 April 2022).

We have proposed a polynomial regression model [37] to address these shortcomings and establish a dynamic binarization threshold prediction model. This model is designed to predict the optimal binarization threshold in real time, enabling it to adapt to changing environmental factors. In this study, we implemented this dynamic binarization threshold prediction model to binarize image data within a specified timeframe. For instance, Figure 8 demonstrates the dynamically predicted threshold values for each hour from 0:00 to 23:00 between 15 March and 18 March 2022.

Using the threshold values predicted by our model, we successfully improved the image recognition accuracy of CCTV images for Chengmei Bridge on the Keelung River spanning 2022 to 2023, as shown in Figure 9.

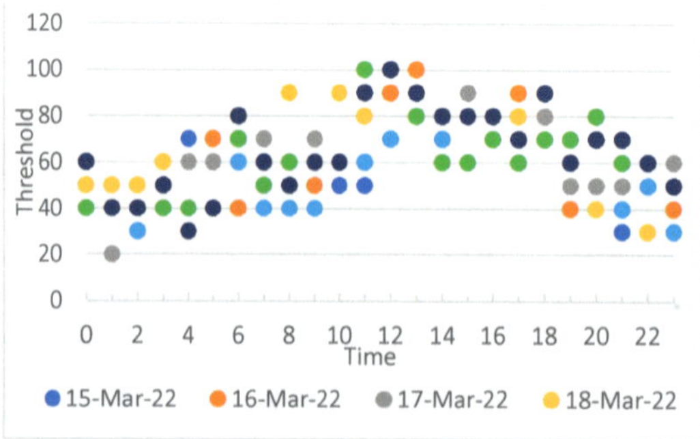


**Figure 8.** Dynamically predicted threshold values for each hour from 0:00 to 23:00 on 15 March to 18 March 2022.

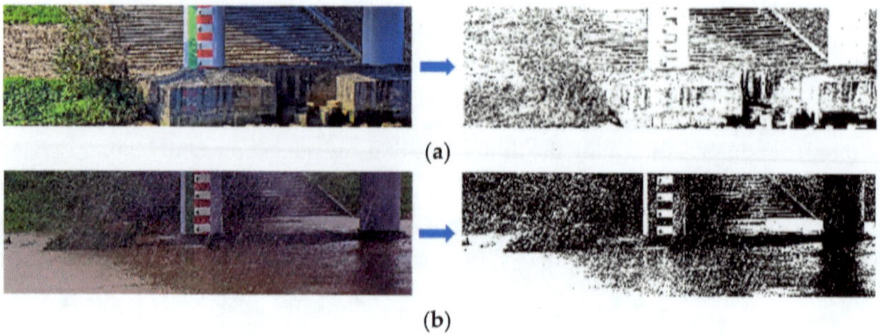

**Figure 9.** (**a**) Binarization of an image captured on a sunny day at 5:00 p.m. with a threshold value of 70. (**b**) Binarization of an image captured on a rainy day at noon with a threshold value of 110. Source: https://fmg.wra.gov.tw/fmgp/ccd_proxy?sn=40 ((**a**) accessed on 7 May 2022, (**b**) accessed on 16 May 2022).

The binarization process of images requires distinct threshold values depending on varying weather conditions and times. For instance, Figure 9a illustrates that a threshold of 70 is optimal for binarizing images taken on a clear, sunny day at 5:00 p.m. Conversely, for images captured on a rainy day at noon, a higher threshold value of 110 is recommended, as demonstrated in Figure 9b.

This dynamic binarization strategy allows for acquiring binarized images ideally suited for recognition tasks, effectively circumventing the constraints of traditional static image binarization. This adaptability to environmental variations ensures reliable and accurate image recognition, regardless of the prevailing conditions during image capture.

### 2.1.5. Selection of Grids for Virtual Water Gauge

This study introduces a grid selection method for creating a virtual water gauge. As illustrated in Figure 10, following the binarization of the image, the process selects an appropriate grid from each row to construct the virtual water gauge. The crucial characteristic of the grids chosen as candidates for the virtual water gauge within the same column is their ability to distinctly differentiate between water-free, partially water-filled, and fully water-filled states. This method enhances the precision in recognizing water levels under varying environmental conditions, thus contributing to advancements in image recognition.

**Figure 10.** Select suitable grids from each row to construct a virtual water gauge.

Figure 11 exemplifies this concept with two images from the same row in their original and binarized forms. Figure 11a depicts a clear day without water, while Figure 11b represents a rainy day with a water-filled grid. It is evident from these images that the blue grid exhibits significant differences in the binarized images between the water-free and fully water-filled states, while the yellow grid does not. Therefore, the blue grid is more suitable than the yellow grid for constructing the virtual water gauge. This innovative method enhances the precision of water level recognition, contributing significantly to advancements in image recognition techniques.

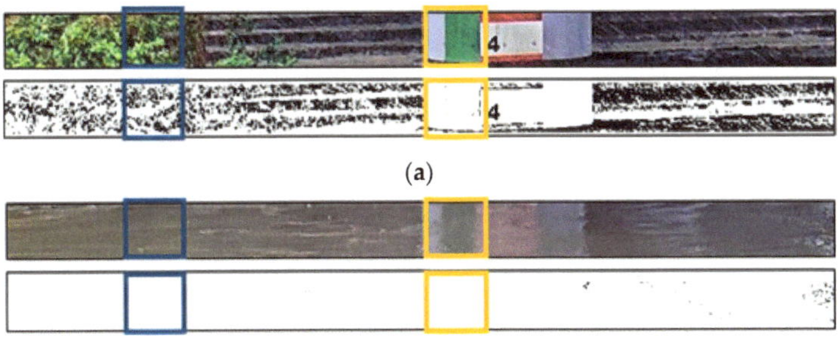

**Figure 11.** Grid selection for the virtual water gauge based on binarized images of different water levels. (**a**) Sunny day with no water in the grid. (**b**) Rainy day with full water in the grid.

The grid selection method proposed in this study encompasses the following four primary steps:

1. Grid Clustering Each identical grid from the collected river images undergoes a binarization process, with all historical data of the same grid categorized into three groups: "no water", "partial water", and "full water", as displayed in Figure 12. To facilitate this process, we employ a semi-supervised auto-labeling technique using ResNet50, and we start with a small set of manually labeled images under the categories of "no water", "partial water", and "full water". Using this labeled set, we train an initial model. This trained model is then used to predict labels for the unlabeled images. Images where the model's predictions are highly confident are identified and added to the labeled set with their predicted labels, known as pseudo-labeling. We then retrain the model on this newly augmented labeled set. This iterative process performs pseudo-labeling and retraining the model until a specified stopping condition is met, such as no significant improvement in the model performance. This semi-supervised auto-labeling approach allows the model to progressively learn from a more extensive dataset, improving its performance, even if the initial labeled dataset is small.
2. Grid Preprocessing Before extracting features with the ResNet50 model, ensuring that the grid size meets the model's input requirements, i.e., 224 × 224, is imperative. If

the grid size does not meet the ResNet50 model's input requirements, the original grid is expanded to 224 × 224 using bilinear interpolation [38–41]. This preserves the details and features of the original grid, ensuring visual consistency and preventing shape or pixel distortion, thereby enabling the extraction of accurate feature vectors.

3. Grid Feature Extraction This step primarily involves extracting features of each grid's "no water", "partial water", and "full water" states using ResNet50. Preprocessed grid historical images are input into the ResNet50 model. Assuming the number of images is n, the input shape is (n, 224, 224). After convolution calculations, the feature vector is extracted from the layer before the fully connected layer, with a shape of (n, 7, 7, 2048), which is then transformed into a one-dimensional array for the subsequent similarity calculations. Figure 13 illustrates that we ultimately obtained the feature vectors for the three categories.
4. Virtual Water Gauge Grid Selection This step aims to select the most suitable grid from each row to serve as a virtual water gauge. Here is the process:
    - Feature Vector Similarity Calculation for Each Grid For each grid, compute its feature vectors under three different water level states: "no water" (denoted as N), "partial water" (denoted as P), and "full water" (denoted as F). These feature vectors, which encapsulate the image characteristics of each grid under different water level states, are obtained through the previously mentioned Grid Feature Extraction model.
    - Cosine Similarity Calculations The cosine similarity formula calculates the similarity between feature vectors of different water level states [42–45], denoted as $S_{A\times B}$. This formula is given by

$$S_{A} \times_{B} = \frac{AB}{||A||\,||B||} \tag{2}$$

where A and B are the feature vectors, • denotes the dot product, and | | A | | and | | B | | are the magnitudes of vectors A and B, respectively. For this analysis, we substitute (A, B) in the formula with (N, P), (N, F), and (P, F) to calculate the cosine similarities for these combinations. A value closer to 1 for a cosine similarity indicates a higher similarity between the feature vectors of two water level states. Therefore, if a grid's feature vectors under different water levels have a high cosine similarity, it implies that the grid's feature vectors are ineffective in distinguishing between water level states, making it less suitable as a virtual water level gauge.

    - Similarity Average Calculations For each grid, we calculate the average of the cosine similarities $S_{N\times P}$, $S_{N\times F}$, and $S_{P\times F}$ to get Similarity[r][k], which represents the similarity of the kth grid in the rth row:

$$\text{Similarity}[r][k] = \frac{(S_{N\times P} + S_{N\times F} + S_{P\times F})}{3} \tag{3}$$

Figure 14 provides a box plot illustrating the similarity values for each row of grids. These values were derived from images captured at the Chengmei Bridge on the Keelung River, collected from 1 March 2022 to 28 February 2023.

    - Select the grid with minimum similarity Finally, we select the grid with the smallest average similarity for each row to serve as the virtual water level gauge. $\text{Grid}[i]_{\text{virtual water gauge}}$ represents the index of the grid with the smallest similarity in the ith row, and argmin_k [46] is a function that delivers the index k of the minimum value in the sequence Similarity[i]:

$$\text{Grid }[i]_{\text{virtual water gauge}} = \text{argmin\_k}(\text{Similarity}[i][k]) \tag{4}$$

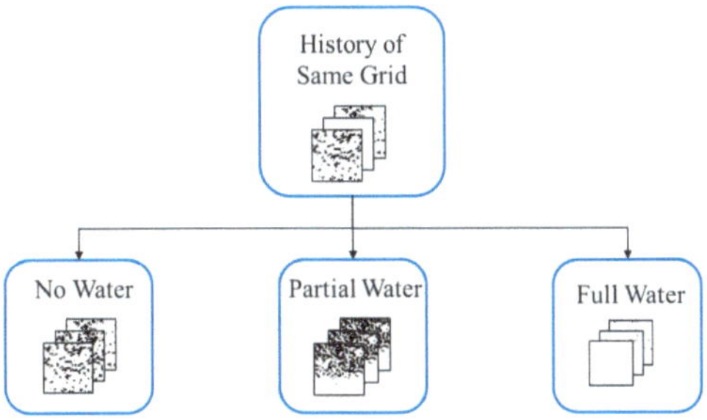


**Figure 12.** Categorization of historical image data of the same grid.

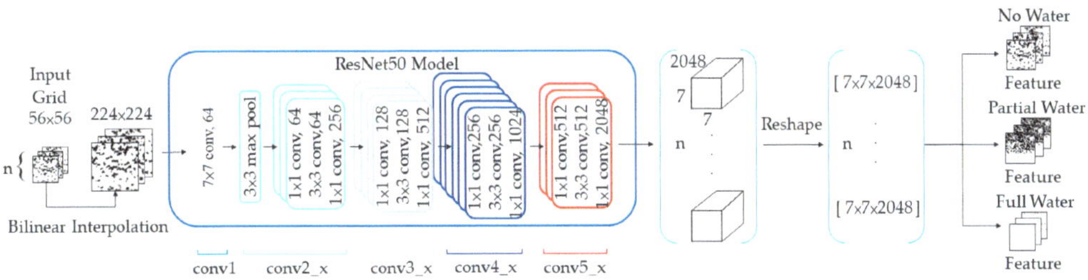


**Figure 13.** Utilizing ResNet50 for grid feature extraction.

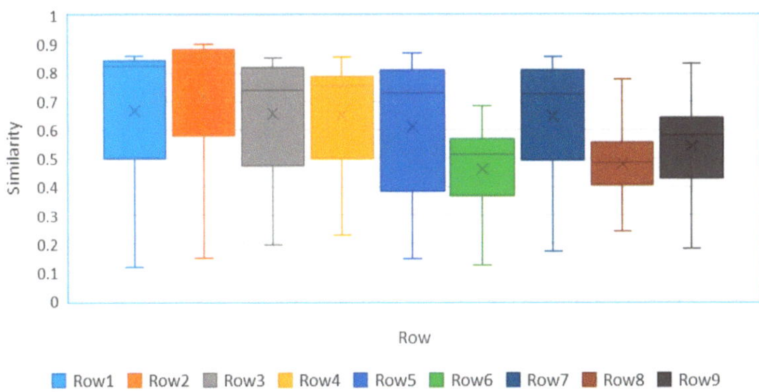


**Figure 14.** Box plot of the similarity values for each row of grids from Chengmei Bridge images.

The steps outlined earlier form the proposed grid selection method in this research. This method aids in identifying the most fitting grid to serve as a virtual water gauge for detecting water levels. As depicted in Figure 15, the grid with the lowest similarity value within its row, marked by a red box, is selected as the virtual water gauge.

### 2.2. Grid State Recognition Model

In this study, we design individual Grid State Recognition Models for each grid identified by the Grid Selection Models to compose the virtual water gauge. The primary purpose of these Grid State Recognition Models is to discern the current water level status of the grid from real-time CCTV footage, categorizing it as "no water", "partial water", or "full water."

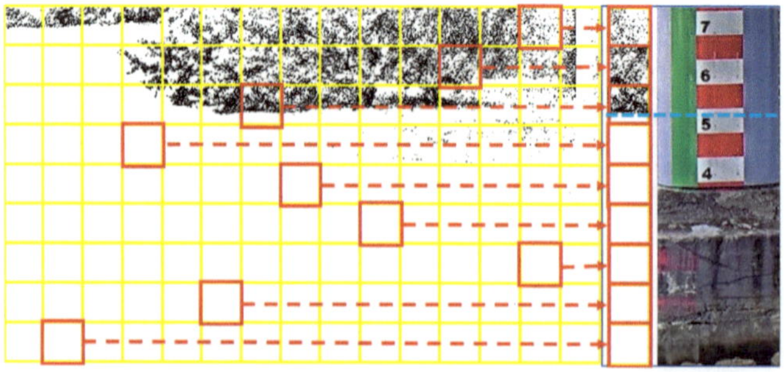

**Figure 15.** Selection of the grid with the lowest similarity in each row as a virtual water gauge. The blue line represents the water level of the virtual water gauge, corresponding to the actual water level height of the physical water gauge.

To build the Grid State Recognition Model, we employ the architecture of the ResNet50 model. The model input for each grid of the virtual water gauge is the binary grid image of all the historical image data after being clustered into the "no water", "partial water", and "full water" categories. We resized the binary grid image to 224 × 224 using bilinear interpolation. After multiple training iterations, we successfully trained a model capable of identifying the grid status as "no water", "partial water", or "full water" based on the input binary grid image, as depicted in Figure 16.

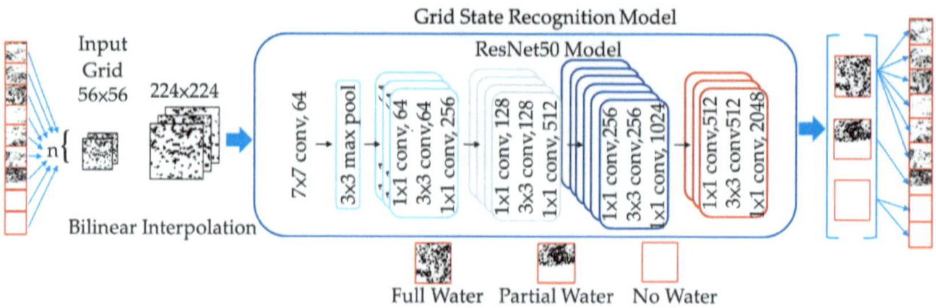


**Figure 16.** Grid State Recognition Model utilizing ResNet50 and bilinear interpolation for determining the water level status in gauge grids.

### 2.3. Water Level Calculation Model

Our Water Level Calculation Model uses the "no water", "partial water", and "full water" states of a virtual water gauge grid, as identified by the Grid State Recognition Model, as the input. These states are utilized to calculate the water level of the virtual water gauge. We assume that the height of a grid corresponds to the actual water level, referred to as $Grid_{Height}$, where $Grid_{Height}$ is equal to 0.86 m based on a grid size of 56 pixels and a correspondence of 1 m to 65 pixels in the river image from the Chengmei Bridge CCTV.

For the "no water" state, the corresponding river water level is 0, while, for the "full water" state, it is $1 \times Grid_{Height}$. For the "partial water" state, we first calculate the grid water level height ΔH within the grid, which is then converted into the corresponding river water level height $\Delta H \times Grid_{Height}$.

The calculation of the water level height ΔH within the "partial water" grid is carried out in four steps, as depicted in Figure 2:

1. Preprocessing The binary "partial water" state is converted by dividing the image values by 255. Pixels with water are converted to 0 and those without water to 1, producing a 56 × 56 matrix.
2. Horizontal Water Ripple Filtering We employ two convolution operations to filter out horizontal water ripples [47–49]. The first convolution operation uses a 1 × 7 filter ma-

trix with all values set to 1 and a stride of 7. Each row undergoes separate convolution operations, resulting in a 56 × 8 matrix (see Figure 17). To calculate the vertical height of the water level in the grid, the 56 × 8 matrix obtained from the first convolution operation undergoes a second convolution operation to produce a 56 × 1 matrix. This operation uses a 3 × 8 filter matrix with all values set to 1 and a stride of 1, with convolution performed from bottom to top. To prevent the original grid height from being affected by the convolution operation, a padding operation [50,51] is performed on the 56 × 8 matrix before the second convolution. The padding matrix is 2 × 8 with all values set to 1, as illustrated in Figure 18.

3. Vertical Water Ripple Filtering For a more precise determination of the water level height, we further process the matrix obtained from the lateral ripple filtering for vertical ripple filtering. A single convolution operation is used to eliminate vertical water ripples gradually. Before this operation, the 56 × 1 matrix undergoes padding, with a 2 × 1 padding matrix where all values are 0. Then, a convolution operation is performed using a 3 × 1 filter matrix with all values set to 1 and a stride of 1. This operation yields a 56 × 1 matrix of water probabilities for each row (see Figure 19).
4. Grid Water Level Height Calculation Lastly we transform the matrix of water probabilities into a matrix of 1s and 0s by setting a threshold value $\theta$. Values below $\theta$ are converted to 0 and those above $\theta$ to 1, resulting in the grid water level height matrix H. The grid water level height $\Delta H$ is then calculated using Equation (5); in this case, $\theta$ is 0.7, and the $Grid_{Pixel}$ is 56.

$$\Delta H = \frac{\sum_{1}^{Grid_{Pixel}} H[i]}{Grid_{Pixel}} \quad (5)$$

Subsequently, the river water level is calculated using Equation (6), where $N_{Full\_Water}$ represents the total number of "full water" grids.

$$Level_{Height} = (1 \times N_{Full\_Water} + \Delta H) \times Gird_{Height}, \quad (6)$$

Our Water Level Calculation Model provides a comprehensive approach for estimating water levels from CCTV images. By integrating the Grid State Recognition model, we can interpret the state of the grid and calculate the corresponding water level. This model is a highly effective tool for real-time monitoring and predicting river water levels, significantly contributing to disaster prevention and management.

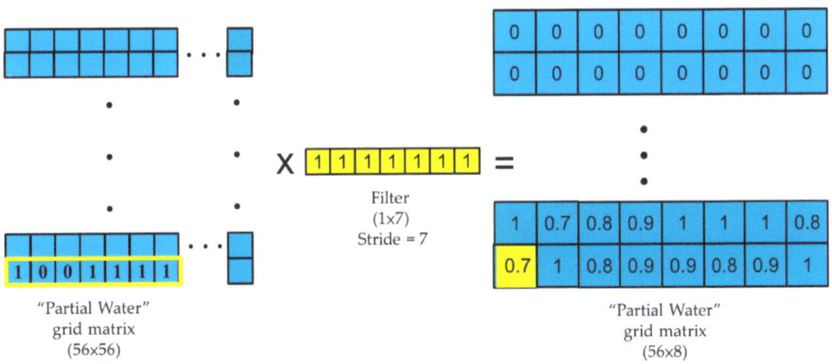


**Figure 17.** Results of convolution operations using a 1 × 7 filter matrix.

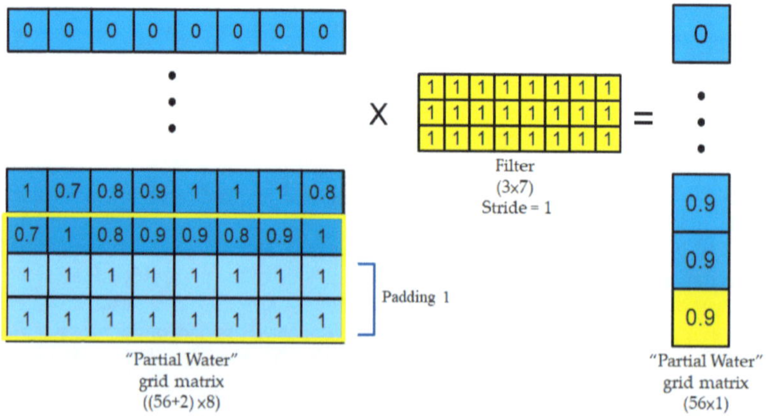


**Figure 18.** Second convolution operation and padding matrix application.

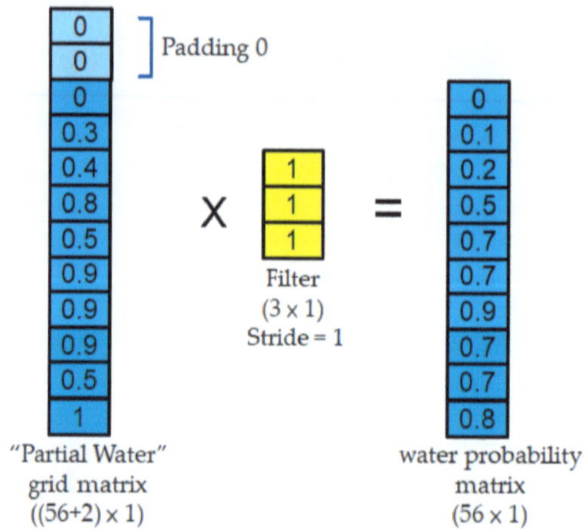


**Figure 19.** Removal of vertical water ripples using a convolution operation.

## 3. Results

This section delves into assessing the model's accuracy, cost-effectiveness, and overall performance. A crucial aspect of the validation procedure entails determining the feasibility of implementing real-time river water level monitoring on a Raspberry Pi platform. The platform from which the study's data are sourced solely provides CCTV river images for the Keelung River Chengmei Bridge, devoid of any physical measurement data. Consequently, our precision assessment relies exclusively on the model's results. Nevertheless, during the preliminary evaluation stages, we manually interpreted the water levels from the images, marking them to establish a baseline for the model assessment. Consequently, we expanded the grid matrix of the virtual water gauge from its original 9 grids, which correspond to the actual water gauge, to 15 grids. This augmentation ensures that the coverage spans the lowermost section of the image.

### 3.1. Model Accuracy

The process of authenticating the model's accuracy principally involved the following essential steps:

1. Image Clustering Leveraging the weather data from the Central Weather Bureau, the river images under investigation were categorized into three distinct classes: sunny, cloudy, and rainy.
2. Initial Water Level Determination The water level for the first frame in the test video was manually annotated to establish an initial reference point.
3. Evaluation of the Virtual Water Gauge Accuracy This process acknowledges that the river water level exhibits sudden surges or drops when an anomaly is observed in the virtual water gauge level, as demonstrated in Figure 20. A rapid decline promptly follows an increase, which we classify as a spike, signifying an error in the water level assessment. Figure 20a,b illustrate the outcomes of the Water Level Calculation Model under rainy and heavy rain conditions, respectively. The results of the Water Level Calculation Model under clear weather conditions are presented in Figure A1. If the current water level diverges from the previous one by more than $\theta$, it is marked as a spike, indicating a potential error in the water level assessment. In this study, $\theta$ was assigned to correspond to $Grid_{Pixel}$. The formula utilized to compute accuracy is provided by Equation (7):

$$\text{Accuracy} = \frac{(\text{Total number of tests} - \text{Number of spikes})}{\text{Total number of tests}} \tag{7}$$

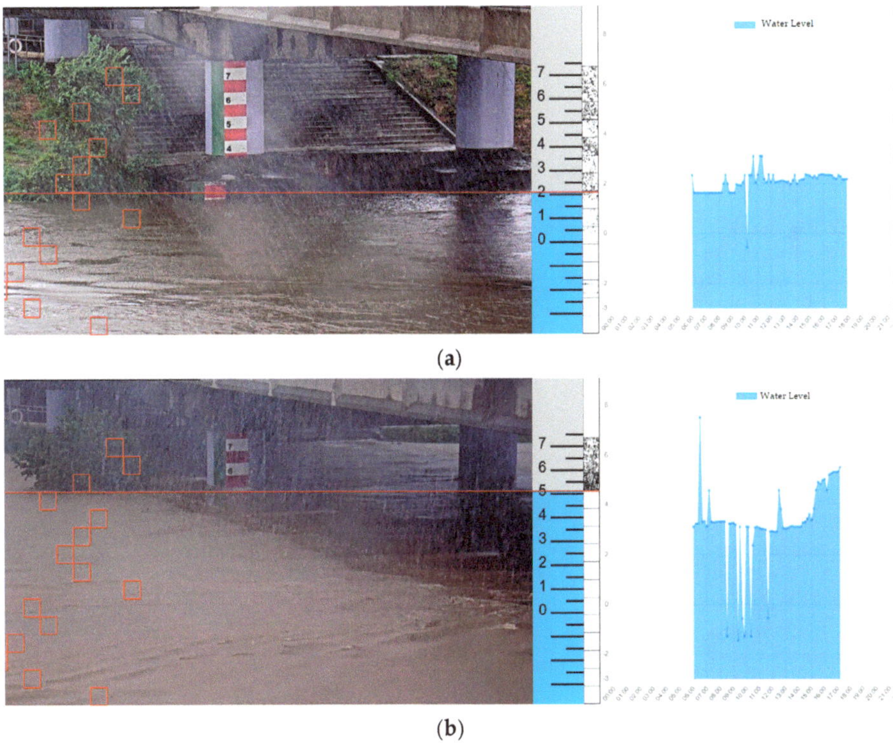

**Figure 20.** The results of the Water Level Calculation Model. (**a**) Rainy, 4 September 2022, 6:00 a.m. to 6:00 p.m. (**b**) Heavy rain, 16 October 2022, 6:00 a.m. to 3:00 p.m. The red squares are the grids constituting the virtual water gauge.

During the initial evaluation, the accuracy of the virtual water gauge level was found to be 93% on sunny days, 88.4% on cloudy days, and a mere 50.1% on rainy days. A thorough analysis of the inaccurate image data indicated that the primary source of these errors was wrong predictions of the dynamic binarization values, which significantly distorted the water level estimates.

Crucial adjustments were implemented to rectify these issues, primarily focusing on refining the dynamic binarization threshold prediction model. However, the original model incorporated weather factors, and the influence of the rainy day factor was reduced due to the limited volume of image data from rainy days.

The dynamic binarization threshold prediction model was restructured into three distinct models, each tailored for sunny, cloudy, and rainy conditions. This modification aimed to bolster the model's adaptability to various weather conditions and, importantly, to equip each model with a dedicated dataset for enhanced learning and prediction accuracy.

The results of these adjustments were highly encouraging. The accuracy of the virtual water gauge level increased to 96.3% on sunny days, 90.9% on cloudy days, and, remarkably, 83.6% on rainy days, as illustrated in Figure 21. These improvements underscore the effectiveness of the refined approach in minimizing errors associated with inaccurate dynamic binarization predictions, particularly under adverse weather conditions.

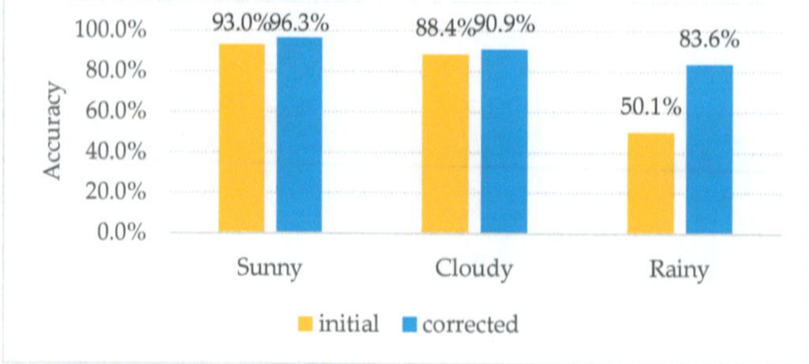


**Figure 21.** Impact of the binarization threshold on the accuracy of the virtual water gauge level.

Upon rectifying the dynamic binarization values, the model demonstrated enhanced accuracy across all weather conditions, with a substantial improvement observed under rainy conditions. This underscores its effectiveness and robustness in accurately determining water levels.

The model's accuracy ranged from 83.6% to 96%. The highest accuracy was recorded during clear weather conditions, while the lowest was observed during heavy rainfall. This fluctuation can be attributed to the complexity and clarity of river imagery under different weather conditions.

### 3.2. Virtual Water Gauge System

This research combines the proposed grid-based virtual water gauge with the Raspberry Pi and corresponding river water level sensor components to design a low-cost system for real-time river water level monitoring.

The system encompasses the following characteristics:

- Main Functions: The CCTV monitoring function can capture images with a resolution of 1920 × 1080. It also detects the image grid water level and calculates the water level height utilizing a virtual water gauge.
- Specifications: The system has sensors for the temperature, humidity, light intensity, and rain detection. It also incorporates a waterproof infrared camera, an Internet of Things communication module, Arduino Mega2560, Raspberry Pi 4B, solar panels, a solar power manager, a rechargeable lithium battery, a waterproof box, and support rods.
- Power Supply: The system can accommodate a maximum power supply of 5 V, 3 A and utilizes a 20 W monocrystalline silicon solar panel and a 10 Ah lithium battery.
- Cost: USD 750.

The developed physical system is depicted in Figure A2. Figure A2a presents an exterior view of the virtual water gauge system, while Figure A2b illustrates the configuration of the components within the waterproof box.

### 3.3. *Grid-Based Methods and Infrastructure Integration*

To verify if the grid-based virtual water gauge system could fulfill the requirements of real-time river water level monitoring using a Raspberry Pi, we implemented the system on both a standard PC and a Raspberry Pi 4 to analyze an image of Chengmei Bridge's water level. The time required for a single frame was recorded as 0.1259 s and 0.5428 s, respectively. Table 1 shows the PC and Raspberry Pi 4B hardware specifications used in this validation process.

**Table 1.** Hardware specifications of the PC and Raspberry Pi 4B used for performance testing.

| Model Type | Specification | Computing Time |
|---|---|---|
| PC | CPU: i5-8300H<br>GPU: GTX: GTX-1650<br>RAM: 20 GB | 0.1259 s |
| Raspberry Pi 4 B | CPU: ARM Cortex A72<br>RAM: 8 GB | 0.5418 s |

We amalgamated grid-based methods with infrastructure elements like CCTV cameras and Raspberry Pi devices for data processing. This fusion of technologies enabled us to devise an efficient system capable of real-time monitoring, thereby enhancing the reliability of our model.

Integrating the virtual water gauge system significantly enhanced the real-time monitoring capabilities of our model. This system, which operates independently of the reliability of physical gauges, demonstrated its ability to provide accurate water level readings while effectively reducing deployment costs.

When applied to real-time river monitoring, this system presents several benefits:

- Speed and Efficiency: As indicated in the tests, the system can process images swiftly, even on a low-powered device like a Raspberry Pi. This facilitates near-real-time monitoring, essential when immediate responses to fluctuating water levels are needed, such as during floods or heavy rainfall.
- Accessibility and Cost-Effectiveness: Using a Raspberry Pi makes the system highly accessible and cost-effective. A Raspberry Pi is affordable and widely available, allowing system deployment in multiple locations without substantial financial expenditure.
- Automation and Accuracy: The system automates the process of water level monitoring, eliminating the need for time-consuming and error-prone manual measurements. Incorporating image recognition and the virtual water gauge also enhances the precision of water level readings.
- Flexibility: The system can be adapted to monitor various rivers or bodies of water by simply changing the image source, making it a flexible solution tailored to diverse monitoring needs.
- Environmentally Friendly: The ability of the system to operate on a low-powered device like a Raspberry Pi means it can be powered by renewable energy sources, such as solar panels, marking it as an environmentally friendly solution.

## 4. Discussion

This section examines the model's accuracy and performance validation, particularly emphasizing its performance under various weather conditions. Furthermore, we investigate how weather conditions affect the precision of the virtual water gauge to enhance the model's decision-making in future scenarios. Despite this, we acknowledge certain constraints of our study, including the necessity to gather data in differing climates and the challenge of establishing the virtual water gauge scale without a physical water gauge. We propose the following solutions to mitigate these challenges and increase the system's practicality.

1. Clarifying Model Accuracy The initial data accuracy for rainy days was 50.1%, mainly due to threshold prediction errors within the dynamic binarization prediction model. This resulted from the reduced sample size during heavy rainfall periods, which diluted the binarization threshold characteristics during the modeling process. While the adjusted accuracy increased to 83.6%, we could further augment the model accuracy by applying techniques such as Resampling [52], Cost-Sensitive Learning [53], or other methods to manage unbalanced datasets and enhance the model's ability to predict binarization thresholds. Alternatively, installing a photometric sensor at the monitoring site to set the binarization threshold directly based on lumen values could minimize errors in the judgment of the virtual water gauge water levels due to the binarization threshold.
2. The Relationship between Weather and Virtual Water Gauge Accuracy We categorized the images into sunny, cloudy, and rainy conditions to analyze the influence of different weather scenarios on the model.
    - During sunny days, spikes were primarily attributed to misinterpretations caused by sun reflections on the grid, fallen leaves or other debris, and large waves created by strong winds on the water's surface.
    - On cloudy days, the reflection and shadow on the water surface, influenced by cloud variations, could cause the intensity of the sunlight on the river surface to change rapidly as the cloud layer moves. This may induce fluctuations in the light intensity during the image processing stage, potentially leading to inaccuracies in water level detection. Consequently, future measures may require algorithm adjustments or the utilization of data from other sensors to mitigate the impact of these light fluctuations on water level measurements.
    - On rainy days, especially during intense rainfall, the selected grid was filled with rainwater, leading to misjudgments of a full water level, or the rain hitting the water surface caused large waves, leading to misjudgments of a waterless state.

    In the future, we can deploy the Boyer–Moore majority algorithm [54,55] or other data mining techniques to identify the primary categories or samples to circumvent or rectify situations where the grid is misjudged.
3. Merits of the Grid-Based Approach The grid-based method used in this study offers two key benefits:
    - Efficiency in Image Processing: With the image size of 1920 × 1080 pixels and the grid size of 56 × 56 pixels, this approach substantially reduces the computational complexity and workload by up to 98.6%. This efficiency enables real-time river water level monitoring, enhancing the system's overall performance.
    - Precision in Height Calculations: Unlike traditional water gauges that use meter units, the grid height calculations allow for a more detailed height scale, capturing exact height data, such as 1.23 m. This granularity provides comprehensive height information, enhancing the accuracy and precision of water level measurements and predictions.
4. Limitations and Future Directions Compared to physical water gauges, a fundamental limitation of this research is the necessity to gather river image data under various climatic conditions for training before deployment unless preexisting historical river image data are available. We plan to introduce an automated process in future research to overcome this limitation. This approach will involve establishing a computerized data collection and processing workflow to save time and ensure data consistency. In instances where rivers are equipped with actual water gauges, the scale of the virtual water gauge can be defined based on the water gauge in the image. However, another challenge arises for rivers without real water gauges. In these cases, a temporary

ruler must be established on-site during the initial setup. This ruler is then screen-captured, and the scale of the virtual water gauge is defined through it. In response to these challenges, our lab is harnessing Augmented Reality (AR) technology to measure object heights using CCTV or mobile camera devices such as mobile phones. This approach aims to alleviate the need for physically establishing a ruler on-site. Furthermore, we plan to apply geometric and trigonometric functions to correct height deviations caused by camera angles. These advancements can significantly enhance the accuracy and applicability of virtual water gauges. By implementing these strategies, we aim to address the current limitations and expand the application of virtual water gauges, enabling swift deployment in most rivers.

This study illuminates the complexities and considerations of utilizing AI technology for intelligent water resource management and environmental sciences. Despite the identified challenges and limitations, this research has demonstrated promising potential for applying AI in this domain. We anticipate that the insights garnered will not only enhance the accuracy and reliability of our existing model but also lay the groundwork for future innovations. As we refine our methodology and explore new techniques, we are committed to advancing our understanding and application of AI technology in hydro-informatics systems toward a more sustainable future.

## 5. Conclusions

The research presented in this paper demonstrates the significant potential of AI technology in environmental sciences and water resource management. Our study focused on developing and validating a grid-based virtual water gauge model, which utilized a Raspberry Pi platform, for real-time river water level monitoring. The model's accuracy varied across different weather conditions, with the best results achieved on sunny days at 96.3%, cloudy days at 90.9%, and rainy days at 83.6%.

This study also culminated in the design of a cost-effective and efficient real-time river water level monitoring system. The system combines the grid-based virtual water gauge with a Raspberry Pi and other corresponding components, providing an accessible and automated solution for water level monitoring. The system demonstrated a strong performance, even on a low-powered device, and offered flexibility in monitoring various rivers or bodies of water.

However, the research also identified certain limitations and challenges. These included data collection under varied climatic conditions and the challenge of establishing a virtual water gauge scale without a physical water gauge. Future work should address these limitations and improve the model's accuracy and robustness.

Potential future improvements include applying unbalanced dataset management techniques to enhance binarization threshold predictions, deploying data mining techniques to rectify grid misjudgments, and automating model training. Additionally, future research could explore applying geometric and trigonometric functions to correct height deviations caused by camera angles.

Despite the challenges, this study provides valuable insights into the application of AI technology in hydro-informatics systems. As we continue to refine our methodology and explore new techniques, we remain committed to enhancing our model's accuracy, reliability, and practicality, laying the groundwork for future innovations in this critical area. Ultimately, we aim to advance the understanding and application of AI technology in environmental sciences, contributing to a more sustainable future.

**Author Contributions:** Conceptualization, J.-F.C., Y.-T.L., and P.-C.W.; methodology, J.-F.C., Y.-T.L., and P.-C.W.; software, Y.-T.L. and P.-C.W.; validation, J.-F.C., Y.-T.L., and P.-C.W.; formal analysis, J.-F.C.; investigation, J.-F.C., Y.-T.L., and P.-C.W.; resources, J.-F.C., Y.-T.L., and P.-C.W.; data curation, J.-F.C., Y.-T.L., and P.-C.W.; writing—original draft preparation, J.-F.C.; writing—review and editing, J.-F.C.; visualization, J.-F.C., Y.-T.L., and P.-C.W.; supervision, J.-F.C.; project administration, J.-F.C.; funding acquisition, J.-F.C. All authors have read and agreed to the published version of the manuscript.

**Funding:** This research was supported by the National Science and Technology Council, Taiwan (Grant number: NSTC 112-2625-M-002-022-).

**Data Availability Statement:** This study utilized images from Chengmei Bridge on the Keelung River, provided by the Water Resources Agency's Hydrologic Cloud Platform, Ministry of Economic Affairs, Taiwan. The data are available at https://fmgb.wra.gov.tw/fmgp (accessed from 1 March 2022 to 28 February 2023). We express our gratitude to the agency for their open data. No new data were created in this study.

**Acknowledgments:** This study is a subproject within a three-year plan to develop low-cost devices for real-time river water level data, benefiting water level forecasting and flood prevention projects.

**Conflicts of Interest:** The authors declare no conflicts of interest.

## Appendix A

**Figure A1.** The results of the Water Level Calculation Model. Sunny, 13 April 2022, 6:00 a.m. to 6:00 p.m. The red squares are the grids constituting the virtual water gauge.

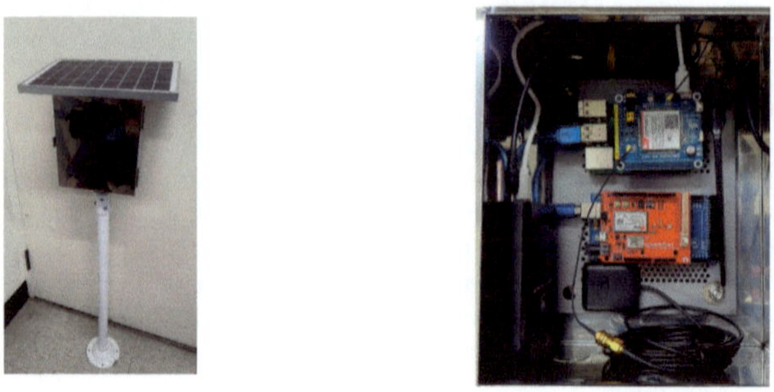

(**a**) (**b**)

**Figure A2.** Virtual water gauge system. (**a**) Physical device. (**b**) The configuration of the components within the waterproof box.

## References

1. Revilla-Romero, B.; Thielen, J.; Salamon, P.; De Groeve, T.; Brakenridge, G.R. Evaluation of the satellite-based Global Flood Detection System for measuring river discharge: Influence of local factors. *Hydrol. Earth Syst. Sci.* **2014**, *18*, 4467–4484. [CrossRef]
2. Alfieri, L.; Bisselink, B.; Dottori, F.; Naumann, G.; De Roo, A.; Salamon, P.; Wyser, K.; Feyen, L. Global projections of river flood risk in a warmer world. *Earth's Future* **2016**, *5*, 171–182. [CrossRef]
3. Alfieri, L.; Dottori, F.; Betts, R.; Salamon, P.; Feyen, L. Multi-Model Projections of River Flood Risk in Europe under Global Warming. *Climate* **2018**, *6*, 6. [CrossRef]
4. Wu, T.; Li, H.-C.; Wei, S.-P.; Chen, W.-B.; Chen, Y.-M.; Su, Y.-F.; Liu, J.-J.; Shih, H.-J. A comprehensive disaster impact assessment of extreme rainfall events under climate change: A case study in Zheng-wen river basin, Taiwan. *Environ. Earth Sci.* **2016**, *75*, 597. [CrossRef]

5. Li, H.-C.; Wei, S.-P.; Cheng, C.-T.; Liou, J.-J.; Chen, Y.-M.; Yeh, K.-C. Applying Risk Analysis to the Disaster Impact of Extreme Typhoon Events Under Climate Change. *J. Disaster Res.* **2015**, *10*, 513–526. [CrossRef]
6. Hsiao, S.-C.; Chiang, W.-S.; Jang, J.-H.; Wu, H.-L.; Lu, W.-S.; Chen, W.-B.; Wu, Y.-T. Flood risk influenced by the compound effect of storm surge and rainfall under climate change for low-lying coastal areas. *Sci. Total Environ.* **2020**, *764*, 144439. [CrossRef]
7. Belabid, N.; Zhao, F.; Brocca, L.; Huang, Y.; Tan, Y. Near-Real-Time Flood Forecasting Based on Satellite Precipitation Products. *Remote Sens.* **2019**, *11*, 252. [CrossRef]
8. Zanchetta, A.D.L.; Coulibaly, P. Recent Advances in Real-Time Pluvial Flash Flood Forecasting. *Water* **2020**, *12*, 570. [CrossRef]
9. Chitwatkulsiri, D.; Miyamoto, H.; Irvine, K.N.; Pilailar, S.; Loc, H.H. Development and Application of a Real-Time Flood Forecasting System (RTFlood System) in a Tropical Urban Area: A Case Study of Ramkhamhaeng Polder, Bangkok, Thailand. *Water* **2022**, *14*, 1641. [CrossRef]
10. Mahapatra, A.; Mahammood, V.; Venkatesh, K. Unsteady flow analysis using hydrological and hydraulic models for real-time flood forecasting in the Vamsadhara river basin. *J. Hydroinform.* **2022**, *24*, 1207–1233. [CrossRef]
11. Tsubaki, R.; Fujita, I.; Tsutsumi, S. Measurement of the flood discharge of a small-sized river using an existing digital video recording system. *J. Hydro-Environ. Res.* **2011**, *5*, 313–321. [CrossRef]
12. Gu, C.; Xu, W.; Wang, G.; Inoue, T.; Rice, J.A.; Ran, L.; Li, C. Noncontact Large-Scale Displacement Tracking: Doppler Radar for Water Level Gauging. *IEEE Microw. Wirel. Compon. Lett.* **2014**, *24*, 899–901. [CrossRef]
13. Zhang, Z.; Zhou, Y.; Liu, H.; Gao, H. In-situ water level measurement using NIR-imaging video camera. *Flow Meas. Instrum.* **2019**, *67*, 95–106. [CrossRef]
14. Chen, G.; Bai, K.; Lin, Z.; Liao, X.; Liu, S.; Lin, Z.; Zhang, Q.; Jia, X. Method on water level ruler reading recognition based on image processing. *Signal Image Video Process.* **2021**, *15*, 33–41. [CrossRef]
15. Zainurin, S.N.; Ismail, W.Z.W.; Mahamud, S.N.I.; Ismail, I.; Jamaludin, J.; Ariffin, K.N.Z.; Kamil, W.M.W.A. Advancements in Monitoring Water Quality Based on Various Sensing Methods: A Systematic Review. *Int. J. Environ. Res. Public Health* **2022**, *19*, 14080. [CrossRef]
16. Yu, J.; Hahn, H. Remote Detection and Monitoring of a Water Level Using Narrow Band Channel. *J. Inf. Sci. Eng.* **2010**, *26*, 71–82.
17. Kim, Y.J.; Park, H.S.; Lee, C.J.; Kim, D.; Seo, M. Development of a cloud-based image water level gauge. *IT Converg. Pract. (INPRA)* **2014**, *2*, 22–29.
18. Hiroi, K.; Kawaguchi, N. FloodEye: Real-time flash flood prediction system for urban complex water flow. In Proceedings of the 2016 IEEE SENSORS, Orlando, FL, USA, 30 October–2 November 2016; pp. 1–3.
19. Pan, J.; Yin, Y.; Xiong, J.; Luo, W.; Gui, G.; Sari, H. Deep learning-based unmanned surveillance systems for observing water levels. *IEEE Access* **2018**, *6*, 73561–73571. [CrossRef]
20. Sabbatini, L.; Palma, L.; Belli, A.; Sini, F.; Pierleoni, P. A Computer Vision System for Staff Gauge in River Flood Monitoring. *Inventions* **2021**, *6*, 79. [CrossRef]
21. Narayanan, R.; Lekshmy, V.M.; Rao, S.; Sasidhar, K. A novel approach to urban flood monitoring using computer vision. In Proceedings of the 2014 5th International Conference on Computing, Communication and Networking Technologies (ICCCNT), Hefei, China, 11–13 July 2014; pp. 1–7. [CrossRef]
22. Chen, M.; Shi, W.; Xie, P.; Silva, V.B.S.; Kousky, V.E.; Higgins, R.W.; Janowiak, J.E. Assessing objective techniques for gauge-based analyses of global daily precipitation. *J. Geophys. Res. Atmos.* **2008**, *113*. [CrossRef]
23. Zhen, Z.; Yang, Z.; Chongzheng, L.; Huabao, L.; Jiabin, Z. Visual detection algorithm of water line based on feature fusion. In Proceedings of the 2019 14th IEEE International Conference on Electronic Measurement & Instruments (ICEMI), Changsha, China, 1–3 November 2019; pp. 474–481. [CrossRef]
24. Yang, F.; Feng, T.; Xu, G.; Chen, Y. Applied method for water-body segmentation based on mask R-CNN. *J. Appl. Remote Sens.* **2020**, *14*, 014502. [CrossRef]
25. Haurum, J.B.; Bahnsen, C.H.; Pedersen, M.; Moeslund, T.B. Water Level Estimation in Sewer Pipes Using Deep Convolutional Neural Networks. *Water* **2020**, *12*, 3412. [CrossRef]
26. Fang, W.; Wang, C.; Chen, X.; Wan, W.; Li, H.; Zhu, S.; Fang, Y.; Liu, B.; Hong, Y. Recognizing Global Reservoirs From Landsat 8 Images: A Deep Learning Approach. *IEEE J. Sel. Top. Appl. Earth Obs. Remote Sens.* **2019**, *12*, 3168–3177. [CrossRef]
27. Pan, M.; Zhou, H.; Cao, J.; Liu, Y.; Hao, J.; Li, S.; Chen, C.-H. Water Level Prediction Model Based on GRU and CNN. *IEEE Access* **2020**, *8*, 60090–60100. [CrossRef]
28. Baek, S.-S.; Pyo, J.; Chun, J.A. Prediction of Water Level and Water Quality Using a CNN-LSTM Combined Deep Learning Approach. *Water* **2020**, *12*, 3399. [CrossRef]
29. Barzegar, R.; Aalami, M.T.; Adamowski, J. Coupling a hybrid CNN-LSTM deep learning model with a Boundary Corrected Maximal Overlap Discrete Wavelet Transform for multiscale Lake water level forecasting. *J. Hydrol.* **2021**, *598*, 126196. [CrossRef]
30. Qiao, G.; Yang, M.; Wang, H. A Water Level Measurement Approach Based on YOLOv5s. *Sensors* **2022**, *22*, 3714. [CrossRef]
31. Xu, Y.; He, C.; Guo, Z.; Chen, Y.; Sun, Y.; Dong, Y. Simulation of Water Level and Flow of Catastrophic Flood Based on the CNN-LSTM Coupling Network. *Water* **2023**, *15*, 2329. [CrossRef]
32. Loizou, K.; Koutroulis, E. Water level sensing: State of the art review and performance evaluation of a low-cost measurement system. *Measurement* **2016**, *89*, 204–214. [CrossRef]
33. Karegar, M.A.; Kusche, J.; Geremia-Nievinski, F.; Larson, K.M. Raspberry Pi Reflector (RPR): A Low-Cost Water-Level Monitoring System Based on GNSS Interferometric Reflectometry. *Water Resour. Res.* **2022**, *58*, e2021WR031713. [CrossRef]

34. Perumal, V.S.A.; Baskaran, K.; Rai, S.K. Implementation of effective and low-cost Building Monitoring System(BMS) using raspberry PI. *Energy Procedia* **2017**, *143*, 179–185. [CrossRef]
35. Kim, I.-K.; Jung, D.-W.; Park, R.-H. Document image binarization based on topographic analysis using a water flow model. *Pattern Recognit.* **2002**, *35*, 265–277. [CrossRef]
36. Dou, G.; Chen, R.; Han, C.; Liu, Z.; Liu, J. Research on Water-Level Recognition Method Based on Image Processing and Convolutional Neural Networks. *Water* **2022**, *14*, 1890. [CrossRef]
37. Chen, J.-F.; Wang, P.-C.; Wong, S.-M.; Liao, Y.-T. Image Recognition of River Water Gauges Using Polynomial Regression Model for Predicting Binarization Threshold. In Proceedings of the 2022 IEEE 4th Eurasia Conference on IOT, Communication and Engineering (ECICE), Yunlin, Taiwan, 28–30 October 2022; pp. 320–324. [CrossRef]
38. Zhang, L.; Li, H.; Zhu, R.; Du, P. An infrared and visible image fusion algorithm based on ResNet-152. *Multimedia Tools Appl.* **2022**, *81*, 9277–9287. [CrossRef]
39. Cha, Y.; Kim, S. The Error-Amended Sharp Edge (EASE) Scheme for Image Zooming. *IEEE Trans. Image Process.* **2007**, *16*, 1496–1505. [CrossRef] [PubMed]
40. Hang, S.T.; Aono, M. Bi-linearly weighted fractional max pooling. *Multimedia Tools Appl.* **2017**, *76*, 22095–22117. [CrossRef]
41. Huang, W.; Xue, Y.; Hu, L.; Liuli, H. S-EEGNet: Electroencephalogram Signal Classification Based on a Separable Convolution Neural Network With Bilinear Interpolation. *IEEE Access* **2020**, *8*, 131636–131646. [CrossRef]
42. Xia, P.; Zhang, L.; Li, F. Learning similarity with cosine similarity ensemble. *Inf. Sci.* **2015**, *307*, 39–52. [CrossRef]
43. Ye, J. Cosine similarity measures for intuitionistic fuzzy sets and their applications. *Math. Comput. Model.* **2011**, *53*, 91–97. [CrossRef]
44. Sejal, D.; Ganeshsingh, T.; Venugopal, K.; Iyengar, S.; Patnaik, L. Image Recommendation Based on ANOVA Cosine Similarity. *Procedia Comput. Sci.* **2016**, *89*, 562–567. [CrossRef]
45. Sadbhawna; Jakhetiya, V.; Chaudhary, S.; Subudhi, B.N.; Lin, W.; Guntuku, S.C. Perceptually Unimportant Information Reduction and Cosine Similarity-Based Quality Assessment of 3D-Synthesized Images. *IEEE Trans. Image Process.* **2022**, *31*, 2027–2039. [CrossRef] [PubMed]
46. Gould, S.; Fernando, B.; Cherian, A.; Anderson, P.; Cruz, R.S.; Guo, E. On differentiating parameterized argmin and argmax problems with application to bi-level optimiza-tion. *arXiv* **2016**, arXiv:1607.05447.
47. Torre, V.; Poggio, T.A. On Edge Detection. *IEEE Trans. Pattern Anal. Mach. Intell.* **1986**, *PAMI-8*, 147–163. [CrossRef] [PubMed]
48. Campillo, F.; Rossi, V. Convolution Particle Filter for Parameter Estimation in General State-Space Models. *IEEE Trans. Aerosp. Electron. Syst.* **2009**, *45*, 1063–1072. [CrossRef]
49. Pang, Y.; Sun, M.; Jiang, X.; Li, X. Convolution in Convolution for Network in Network. *IEEE Trans. Neural Netw. Learn. Syst.* **2017**, *29*, 1587–1597. [CrossRef]
50. Liu, G.; Dundar, A.; Shih, K.J.; Wang, T.-C.; Reda, F.A.; Sapra, K.; Yu, Z.; Yang, X.; Tao, A.; Catanzaro, B. Partial Convolution for Padding, Inpainting, and Image Synthesis. *IEEE Trans. Pattern Anal. Mach. Intell.* **2022**, *45*, 6096–6110. [CrossRef]
51. Hashemi, M. Enlarging smaller images before inputting into convolutional neural network: Zero-padding vs. interpolation. *J. Big Data* **2019**, *6*, 98. [CrossRef]
52. Yu, C.H. Resampling methods: Concepts, Applications, and Justification. *Pract. Assess. Res. Eval.* **2019**, *8*, 19. [CrossRef]
53. Zadrozny, B.; Langford, J.; Abe, N. Cost-sensitive learning by cost-proportionate example weighting. In Proceedings of the Third IEEE International Conference on Data Mining, Melbourne, FL, USA, 19–22 November 2003; Institute of Electrical and Electronics Engineers (IEEE): Melbourne, FL, USA, 2004. [CrossRef]
54. Boyer, R.; Kaufmann, M.; Moore, J. The Boyer-Moore theorem prover and its interactive enhancement. *Comput. Math. Appl.* **1995**, *29*, 27–62. [CrossRef]
55. Jang, S.; Lim, H. AsyncFL: Asynchronous Federated Learning Using Majority Voting with Quantized Model Updates (Student Abstract). *Proc. AAAI Conf. Artif. Intell.* **2022**, *36*, 12975–12976. [CrossRef]

*water*

MDPI

*Article*

# Research on Flood Risk Control Methods and Reservoir Flood Control Operation Oriented towards Floodwater Utilization

**Liwei Zhou [1,2], Ling Kang [1,2,*], Shuai Hou [1] and Jinlei Guo [1]**

[1] School of Civil and Hydraulic Engineering, Huazhong University of Science and Technology, Wuhan 430074, China; zhoulw@hust.edu.cn (L.Z.)
[2] Hubei Key Laboratory of Digital Valley Science and Technology, Wuhan 430074, China
* Correspondence: kling@hust.edu.cn

**Abstract:** Since improving floodwater utilization may increase flood risk, flood risk control methods for trade-offs between these factors have research value. This study presented a flood risk control method oriented towards floodwater utilization which considers multiple main flood risk factors. The proposed method not only achieves the boundaries of the flood limited water level (FLWL) under various acceptable risks but also dynamically controls the water level to enhance floodwater utilization. A case study conducted on the Danjiangkou reservoir yielded the following results: (1) The proposed method provides FLWL dynamic control boundaries under various acceptable risks. (2) The proposed method reveals the potential to raise the FLWL, with a possibility to raise it by 1.00 m above the present FLWL under the absence of flood risk. (3) The available flood resources in both the wet and dry seasons increase, on average, by 0.83 and 0.81 billion m$^3$, and the flood risk remains within the acceptable range after raising the FLWL by 1.00 m, which contributes to enhancing floodwater utilization.

**Keywords:** dynamic control boundary; risk analysis; flood risk control map; floodwater utilization; acceptable flood risk

## 1. Introduction

Floods are significant natural disasters that result in casualties and property damage [1]. Reservoirs are crucial for flood control and floodwater utilization and lead to conflicts between the two [2]. With growing populations and socio-economic development, water shortages in some regions are becoming more acute [3]. Therefore, enhancing floodwater utilization becomes necessary [4]. The flood limit water level (FLWL) balances flood control and floodwater utilization [5]. In China, reservoirs typically stay below the FLWL during the wet season [6]. However, the conventional FLWL is designed to overemphasize low-probability floods, leading to the insufficient utilization of floodwater [7]. Realizing the dynamic control of the flood limit water level (DC-FLWL) is an available way to improve floodwater utilization; this involves controlling water levels in safe regions for trade-off benefits between flood risk and floodwater utilization [8].

The relationship between risks and benefits is characterized by mutual antagonism, and enhancing the advantages will inevitably entail certain potential hazards. Realizing the dynamic control of the FLWL enhances floodwater utilization, yet it also introduces uncertainties to a certain extent, potentially giving rise to additional flood-related risks [9]. Therefore, research on the DC-FLWL and the generated risks associated with this approach has become popular in recent years. Tan et al. [10] studied the DC-FLWL taking into account the spatial uncertainty of floods. Zhou et al. [11] realized the DC-FLWL with the aim of improving water resource utilization using a multi-objective optimization algorithm and an aggregation decomposition method. Zhang et al. [12] discussed and improved the DC-FLWL on the basis of forecast information. Gong et al. [13] conceptualized river flood routing as a hypothetical reservoir based on the Muskingum model and analyzed its impact on dynamic control boundaries. Ning et al. [14] assessed flood risk by considering flood forecast uncertainty and analyzed the impacts at different FLWLs. Mu et al. [15] established the dynamic water level to optimize water resource utilization and estimated the flood risk resulting from flood forecast errors. Pan et al. [16] identified dynamic control areas using the pre-discharge method and built a risk analysis model. Lu et al. [17] integrated a flood risk analysis model with various risk sources and studied their impacts on flood risk. Du et al. [18] discovered both the risks and benefits of floodwater utilization increases in the wake of increasing the FLWL, albeit at different rates.

The theories and methods for realizing the dynamic control of FLWL and flood risk control are being continuously enhanced. Nevertheless, there are still unresolved issues that warrant further investigation in the field of the DC-FLWL and its associated risks: (1) The existing research on the DC-FLWL mainly focuses on studying the dynamic control boundaries and does not consider the variability of the dynamic control boundaries under different acceptable flood risks. (2) Previous studies have primarily concentrated on floodwater utilization while paying limited attention to reservoir operation and risk control in the context of the DC-FLWL. The DC-FLWL is crucial to improving floodwater utilization and ensuring an acceptable flood risk, which is also an urgent problem in the current research on the flood risk control method. This paper identified key risk factors and proposed a flood risk control method oriented towards floodwater utilization. The proposed method achieved the dynamic control boundaries of the FLWL under various acceptable risks and provided a supportive role for the DC-FLWL.

## 2. Methodology

The risk control method oriented towards floodwater utilization proposed in this paper consists of the following components: (1) The identification of the main risk factors affecting flood control operation; (2) the uncertainty of the main risk factors, including reservoir inflow, interval floods, and forecast errors; (3) the development of a risk analysis model, (4) the modeling of the flood risk control method; (5) an assessment of the risks and benefits of floodwater utilization.

### *2.1. The Identification of The Main Flood Risk Factors*

There are various flood risk factors that can affect reservoir flood control operation, including flood shape, flood forecast error, outflow discharge error, scheduling lag time, and river flood routing errors [19]. It is not feasible to analyze the combined influence of all risk factors. Instead, it is important to consider the main risk factors comprehensively [20]. The forecasted flood is a significant source of uncertainty and serves as the primary input for the model. Additionally, the uncertainties of interval floods must be also considered for reservoirs with downstream flood control tasks [21]. Therefore, the uncertainties of reservoir inflow, interval floods, and flood forecast errors are key focal points in the following paragraphs.

### *2.2. An Uncertainty Analysis of The Main Flood Risk Factors*

#### 2.2.1. The Uncertainty of Reservoir Inflows

Since the uncertainty of reservoir inflows is one of the most important risk factors, it is crucial to have a large number of reservoir inflows that encompass different types, magnitudes, and shapes. While design floods and historical floods provide some insight, they are not sufficient. Therefore, it is necessary to simulate reservoir inflows [22]. Disaggregation methods break down runoff from larger time scales into smaller ones based on historical runoff, which effectively captures the statistical characteristics of both large- and small-scale runoff [23].

In this study, the correlated disaggregation model was used due to its ease of implementation and the availability of code for this model. Taking the flood volume (over $T$

days) during the wet season decomposed into daily flow components as an example, the basic form of the correlated disaggregation model is shown as the following equation:

$$Q = AV + B\varepsilon \tag{1}$$

where $\boldsymbol{Q}$ is the daily flow during the wet season; $\boldsymbol{V}$ is the flood volume during the wet season, which is generally believed to follow the Log-Person Type III distribution; and $\boldsymbol{A}$, $\boldsymbol{B}$ and $\varepsilon$ are the parameter matrixes of the model. The parameter estimation and simulation steps of the correlated disaggregation model are mentioned in [24].

2.2.2. The Uncertainty of Interval Floods

For a reservoir with a downstream area flood control task, the uncertainty of the interval flood is a significant flood risk factor [25]. When interval floods do not meet the requirements of flood risk analysis, alternatives can be generated in accordance with the joint distribution relationship between the reservoir inflow and the interval flood [26]. Their joint distribution relationship is a complex multivariate issue that can be built by using the Copula function.

The Copula function describes the correlation between multiple variables which can connect their marginal distributions with their joint distribution when their marginal distributions have already been determined with certainty [27]. The joint distribution function can be represented as the following equation:

$$G(x_1, x_2) = C_\theta(M_1(x_1), M_2(x_2)) \tag{2}$$

where $G(x_1,x_2)$ is the joint distribution function; $M_1(x_1)$ and $M_2(x_2)$ are the marginal distribution functions of $X_1$ and $X_2$, respectively; and $C_\theta(\cdot)$ is the Copula function. The Copula function and its parameters are optimized based on historical runoff, and the optimization process is discussed in [28].

2.2.3. The Uncertainty of Forecast Error

The forecast error is a significant factor contributing to the uncertainty in flood risk. Assuming an unbiased flood forecast, the forecasted flow at each time follows normal distribution [29]. The relative error is used to comprehensively reflect the forecast error, and the following equation can be obtained:

$$U(t) = O(t)[1 + \alpha(t)] \tag{3}$$

where $U(t)$ is the forecasted inflow or interval flood; $O(t)$ is the observed inflow or interval flood; and $\alpha(t)$ is the relative error. Both the relative errors of inflow and interval flood obey the normal distribution, and their distribution parameters can be estimated in accordance with historical flood forecast results [30].

*2.3. The Development of a Risk Analysis Model*

This paper analyzes the flood risk generated by the main risk factors described in the previous section. The Monte Carlo simulation method [31] was performed $m$ times on the basis of operation rules. When the flood risk control objective is compromised under $n$ floods, flood risk ratio (RR) can be mathematically described as Equation (4):

$$RR = P(A) = P\Big(Z(t) > Z_{\max}\Big|\Big|D(t) > D_{\text{safety}}\Big) = \frac{n}{m} \times 100\% \tag{4}$$

where $P(A)$ denotes the probability of damage to the risk control objective, which is defined by the water level not exceeding the flood control water level (FCWL) and the discharge of the flood control station not exceeding the allowable discharge shown in this paper; $Z(t)$ is the water stage; $D(t)$ is the discharge of the flood control station; $Z_{\max}$ is the FCWL; and $D_{\text{safety}}$ is the allowable discharge of the flood control station.

### *2.4. Flood Risk Control Method Oriented towards Floodwater Utilization*

In this paper, the upper limit of the FLWL is taken for the risk control water level (RCWL), and the lower boundary of FLWL is the present FLWL. Therefore, when RCWLs under various flood return periods are obtained by using the reservoir flood control risk analysis model, dynamic control domain under a certain acceptable risk ratio can be realized. Figure 1 provides a schematic illustration of how one calculates the RCWL, and the detailed process is explained below.

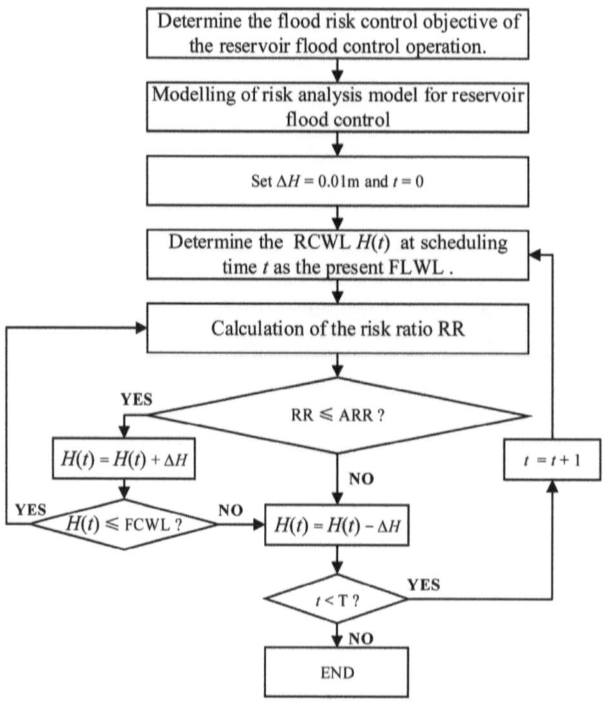


**Figure 1.** Schematic illustration detailing the calculation of the RCWL.

Step 1: Define the flood risk control objective. The water level does not exceed the FCWL, and the discharge of the flood control station does not exceed its allowable discharge.

Step 2: Develop the risk analysis model. The risk analysis model is developed on the basis of the flood risk control objective and operation rules.

Step 3: Determine the computational accuracy of the RCWL as $\Delta H = 0.01$ m, and set the scheduling time as $t = 0$.

Step 4: Set the present RCWL $H(t)$ at scheduling time $t$ as the present FLWL.

Step 5: Calculate the risk ratio (RR) using the risk analysis model.

Step 6: If RR is not greater than the acceptable risk ratio (ARR), set $H(t) = H(t) + \Delta H$ and switch to Step 7. Otherwise, set $H(t) = H(t) - \Delta H$ and switch to Step 8.

Step 7: If $H(t)$ is not higher than the FCWL, return to Step 5. Otherwise, set $H(t) = H(t) - \Delta H$ and switch to Step 8.

Step 8: If $t$ is less than $T$, which represents the number of scheduling periods, set $t = t + 1$ and execute Step 4. Elsewise, finish this process.

For this paper, the RCWLs and the boundaries of the FLWL under various acceptable risks were determined through extensive simulations of floods. Subsequently, the reservoir flood risk control map was generated by considering the lower boundaries of the FLWL. For the wet season, the flood risk control map was directly applied to the DC-FLWL. If the water level exceeds the upper boundary under acceptable risk, the risk will be greater than the acceptable risk. In such cases, the water level should be decreased below the dynamic control upper boundary of FLWL to reduce flood risk based on forecast information. Conversely, the water level should be elevated within the dynamic control domain of the FLWL to enhance floodwater utilization based on forecast information.

Therefore, the core principle of the flood risk control method is realizing the DC-FLWL for balancing floodwater utilization and flood risk. Based on a reservoir flood risk control map, the risk control method in this paper, facilitates improved floodwater utilization within acceptable flood risks, offering technical support for reducing flood risk and enhancing floodwater utilization in reservoir flood control operation.

### 2.5. Risk and Benefit Assessment

As mentioned above, the risks and benefits of floodwater utilization are influenced by the water level, and elevated water levels increase flood risk but enhance floodwater utilization. In this paper, the highest water level (HWL), the average storage capacity deducted from the dead storage capacity during the wet season (ASC), and the storage capacity deducted from the dead storage capacity in the last stage of the wet season (SCT) are used to evaluate the risks and benefits of floodwater utilization.

$$HWL = \max_{t\in[1,T]}\{Z_t\} \tag{5}$$

$$ASC = \frac{\sum\limits_{t=1}^{T}(V_t - V_{DWL})}{T} \tag{6}$$

$$SCT = V_T - V_{DWL} \tag{7}$$

where $V_t$ and $Z_t$ are the storage capacity and water level, respectively; $V_T$ is the storage capacity in the last stage of the wet season; and $V_{DWL}$ is the dead storage capacity. The HWL is used to measure the flood risk, and the ASC and the SCT are both used to measure the benefits of floodwater utilization.

## 3. Case Study

The Danjiangkou reservoir in China's Han river is a water source project for the middle route of the South-to-North Water Diversion Project, the aim of which is to alleviate the water scarcity of 19 large and medium-sized cities [32]. The first phase of this water diversion project created a multi-year average annual transfer of 9.5 billion $m^3$ of water. As of 30 March 2023, this water diversion project has transferred over 55 billion $m^3$ of water and implemented approximately 9 billion $m^3$ of ecological water replenishment. This has directly benefited more than 85 million people, including over 5 million individuals who no longer have to deal with bitter and salty water or highly fluoridated water. The Danjiangkou reservoir is a multi-year adjustment water conservancy project, which provides various benefits. On the basis of the optimal scheduling plan of the Danjiangkou reservoir, it carries out flood control for the Han River, with the Huangzhuang station serving as the control station. The wet season spans from 21 June to 10 October. During the summer wet season, which occurs before 20 August, the FLWL is set at 160.00 m. Similarly, during the autumn wet season, which starts from 1 September, the FLWL remains at 163.50 m. Figure 2 plots the geographical location relationship between the Danjiangkou reservoir and the Huangzhuang station.

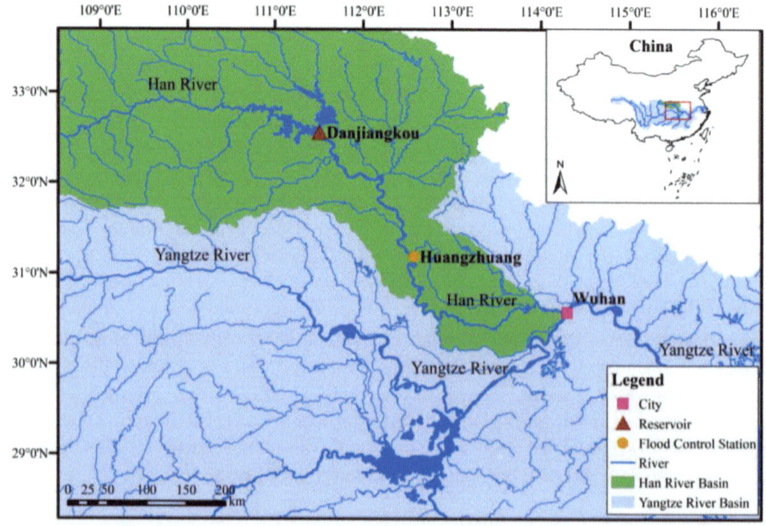


**Figure 2.** The geographical location relationship between the Danjiangkou reservoir and Huangzhuang station.

According to the optimal scheduling plan of the Danjiangkou reservoir, the protection standard for the downstream control area is set at a frequency of once in 100 years. For the sake of guaranteeing downstream safety for floods that occur once in 100 years or less, the reservoir operates using a graded compensation and regulation mode. For floods that exceed the rate of occurring once in 100 years, the reservoir operates using a graded control and discharge mode. Therefore, this paper only considers the floodwater utilization operation mode of the Danjiangkou reservoir for floods that occur once in 100 years or less.

The Danjiangkou reservoir operates according to the proposed operation rules during the wet season and maintains the water level near the FLWL. However, there is a tendency to release water during the wet season, and it is difficult to refill the reservoir in the last stage of wet season, resulting in the prominent conflict between water supply and flood control. This paper considers four historical floods (ones in 1935 and 1975 in the summer wet season and ones in 1964 and 1983 in the autumn wet season) and upscales them to design floods with magnitudes of once in every 5, 10, 20, 50, and 100 years. Additionally, historical daily runoff data from 1969 to 2018 (50 years) and the interval historical daily runoff data from the Danjiangkou reservoir and Huangzhuang station from 1989 to 2018 (30 years) during the wet season were used for the work presented in this paper.

## 4. Results

### 4.1. Simulation of Reservoir Inflow

The correlated disaggregation model was constructed based the on Danjiangkou reservoir daily inflow during the wet season from 1969 to 2018 and evaluated by using statistics, including the mean, coefficient of variation, daily lag-1 autocorrelation, and probability density function. Taking the summer wet season as an example, the daily reservoir inflow processes of the last 50 years were simulated 100 times. Box plots of statistics pertaining to the simulated daily inflow are shown in Figure 3. The light blue boxes describe the distribution characteristics of the corresponding statistic, while the horizontal line represents the median of the corresponding statistic. The red dots connected by a red solid line describe the corresponding statistic of the historical data. As shown in Figure 3, the red dots are all inside the light blue boxes, which indicates that the obtained daily inflow is consistent with the statistics of historical daily inflow. In addition, the statistics of the simulated data exhibit a similar trend to that of the historical data, which indicates that the statistical properties of the historical daily inflow have been well captured and that the simulated daily inflow exhibits the same continuity as the historical data.

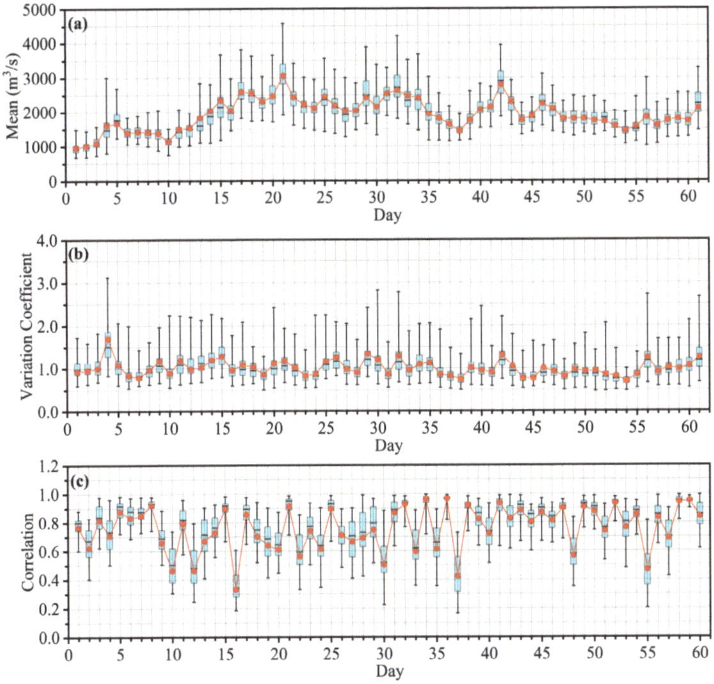


**Figure 3.** Box plots of the statistics for simulated daily inflow. (**a**) The mean; (**b**) the coefficient of variation; (**c**) the daily lag-1 autocorrelation.

In order to further verify that the correlated disaggregation model can effectively reproduce the observed statistics of flood volume (FV), the FV in 1 day, 3 days, 7 days, and 15 days of the simulated and observed inflow were calculated, respectively. Box plots of the probability density functions of the simulated FV1, FV3, FV7, and FV15 are shown in Figure 4. It can be seen that the historical probability density functions of FV, which are highly skewed, are very well captured in different time periods. Therefore, the correlated disaggregation model restores the entire distribution features of FV and daily inflow and can be used to simulate the mass of the daily inflow of the Danjiangkou reservoir.

### 4.2. Simulation of Interval Flood

The reservoir inflow floods and the interval floods during the wet season from 1989 to 2018 were used to establish their joint distribution relationship. Taking the summer wet season as an example, the contour plot of joint distribution probabilities between reservoir flood volume and interval flood volume is illustrated in Figure 5. Their joint distribution probability can be conveniently obtained from Figure 5; if the reservoir flood volume and the interval flood volume are greater than 13.5 and 12.5 billion m$^3$, respectively, their joint distribution probability is about 1% (see the red auxiliary line). Once the joint distribution relationship between reservoir flood volume and interval flood volume was obtained, on the basis of the mass of the simulated reservoir floods, it is possible to obtain the abundant interval floods by using the same ratio amplification method according to the 7-day flood volume under a certain flood return period.

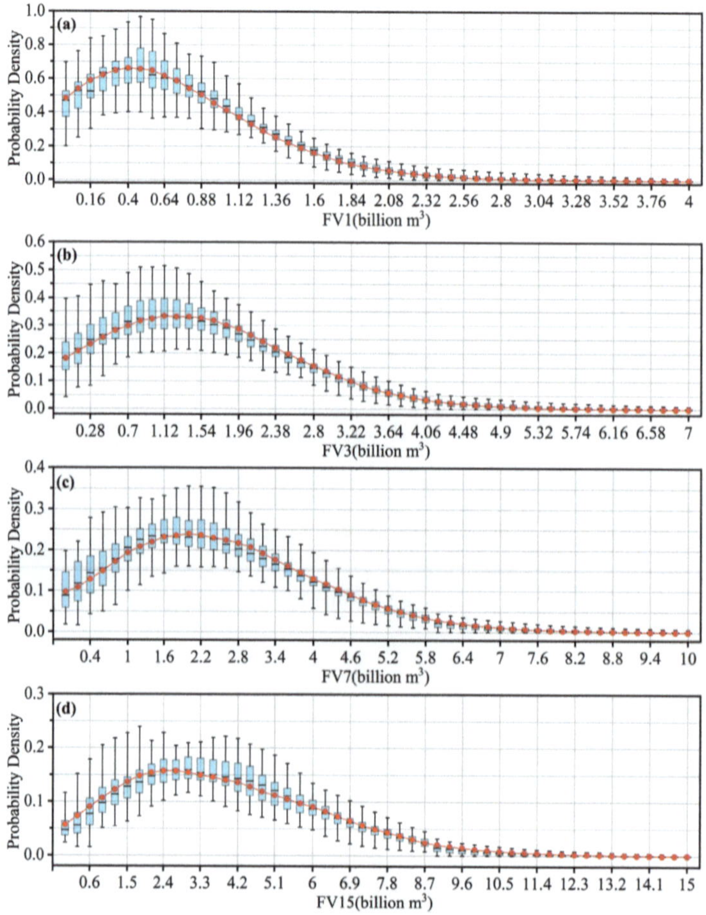


**Figure 4.** Box plots of the probability density functions of simulated FVs in different time periods. (**a**) FV1; (**b**) FV3; (**c**) FV7; (**d**) FV15.

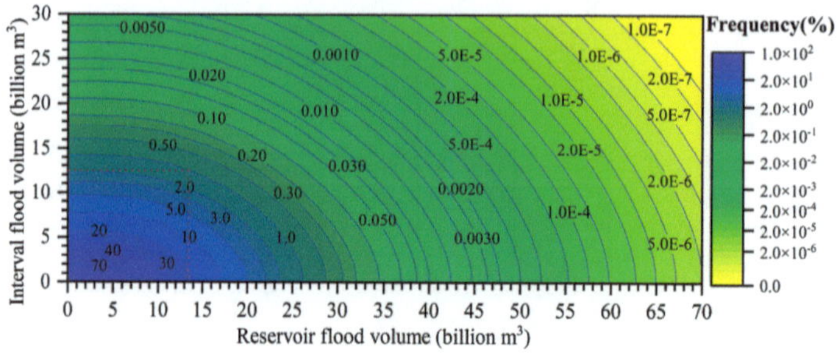


**Figure 5.** Contour plot of joint distribution probabilities between reservoir flood volume and interval flood volume.

*4.3. The Influence of Flood Forecast Error*

According to the "Danjiangkou Water Conservancy Project Flood Control Operation Special Report", the maximum relative error of the reservoir flood forecast $\alpha_I$ and the interval flood forecast $\alpha_q$ are 30% and 20%, respectively. Taking into account the most unfavorable conditions, the forecasted reservoir flood values were assumed to be smaller than the observed values, while the forecasted interval flood values were assumed to be larger than the observed values. To study the influence of flood forecast error, the relative errors were deemed to change in interval of 5%. Taking the once-in-100-year flood of 1975's summer wet season as an example, each set of relative errors was randomly generated for 10,000 forecast floods. Statistics for the highest water level values under various relative errors are shown in Figure 6.

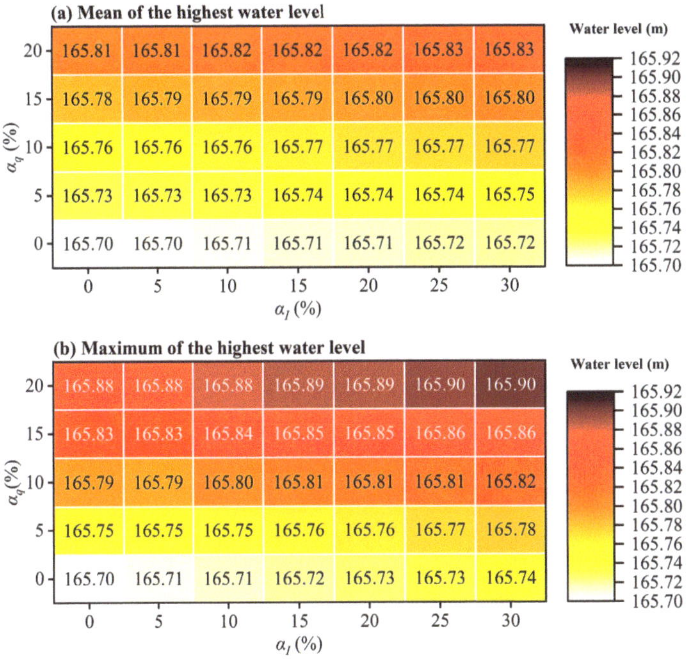


**Figure 6.** Statistics for the highest water level values under various relative errors. (**a**) The mean; (**b**) the maximum.

Figure 6 reveals that the mean and maximum of highest water level gradually increases with relative error. There is no flood forecast error when $\alpha_I$ = 0% and $\alpha_q$ = 0%, and the mean and maximum are both 165.70 m. The larger the relative error is, the larger the mean and maximum are and the higher the flood risk may be.

### *4.4. A Flood Risk Control Map and Flood Risk Control Method Oriented toward Floodwater Utilization*

The RCWLs of the reservoir for various acceptable risk ratios (0, 5%, 10%, 15%, 20%) and flood return periods (5a, 10a, 20a, 50a, 100a) can be calculated by calculating the RCWL based on 30,000 simulated floods. The RCWLs for the same acceptable risk ratio and various flood return periods can be used to draw the dynamic control domain of the FLWL under an acceptable risk ratio. In the absence of flood risk, diagrams of the RCWLs for various flood return periods are shown in Figure 7. The light blue area represents the dynamic control domain of the FLWL, situated above current the FLWL (160.00 m and 163.50 m, respectively) and below the lower boundary line of the RCWLs for various flood return periods. The reservoir operates without flood risk by dynamically controlling the water level within the light blue area.

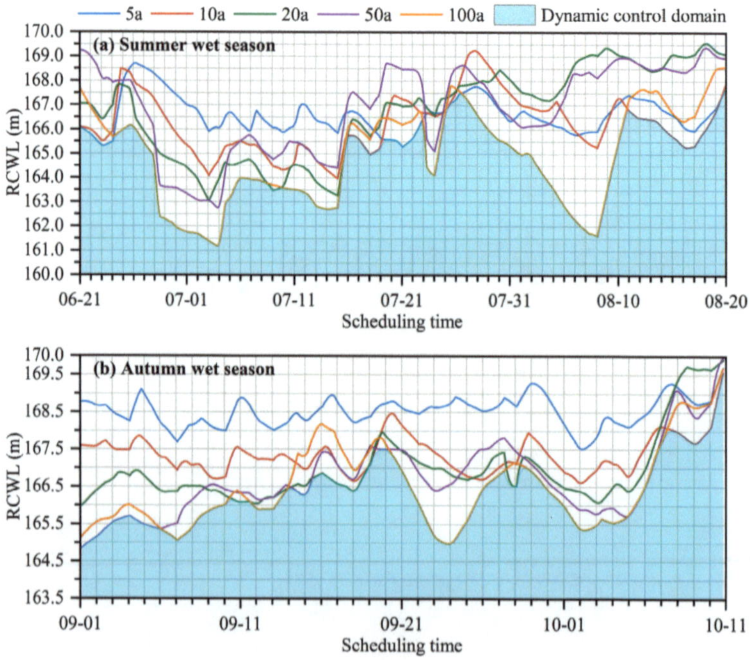


**Figure 7.** Diagrams reflecting the RCWLs for various flood return periods. (**a**) The summer wet season; (**b**) the autumn wet season.

The dynamic control domains of the FLWL under various acceptable risk ratios can be used to draw flood risk control maps, as proven by Figure 8. The dynamic control domain under an acceptable risk ratio and its corresponding flood risk ratio can be easily determined from Figure 8. For example, if the acceptable risk ratio is 5%, the water level should be dynamically controlled in the green area. Flood risk control maps can be also used to study the potential of raising the FLWL under the acceptable risk ratio. Assuming an acceptable risk ratio of 0 (under the absence of flood risk scenario), the minimum RCWLs during the summer and autumn wet seasons are 161.13 m and 164.86 m (see Figures 7 and 8), respectively, and raising the FLWL by 1.00 m based on the current FLWL poses no flood risk for the four historical floods upscaled to various flood return periods. Therefore, from a flood control safety perspective, it can be concluded that there is no flood risk when the FLWLs of the reservoir are raised to 161.00 m and 164.50 m (an increase of 1.00 m from the current FLWL), respectively, during the summer and autumn wet seasons.

*4.5. The Risks and Benefits of Floodwater Utilization*

Data pertaining to a total of 50 years (1969–2018) of the historical floods during the wet season were used to evaluate the risks and benefits of floodwater utilization after the FLWL is raised by 1.00 m, and the findings indicate that the discharge of the Huangzhuang station did not exceed its allowable limit in any of these years. The changes in the HWL, ASC, and SCT before and after the FLWL was raised are noted in Figure 9.

Figure 9a reveals that the HWL increases after FLWL is raised. The annual mean and maximum of the highest HWLs after the FLWL was raised are 165.24 m and 170.59 m, respectively, which are larger than the values before the FLWL was raised. Nonetheless, the HWL after the FLWL raised does not exceed the FCWL (171.70 m), indicating that the flood risk remains within a controllable range.

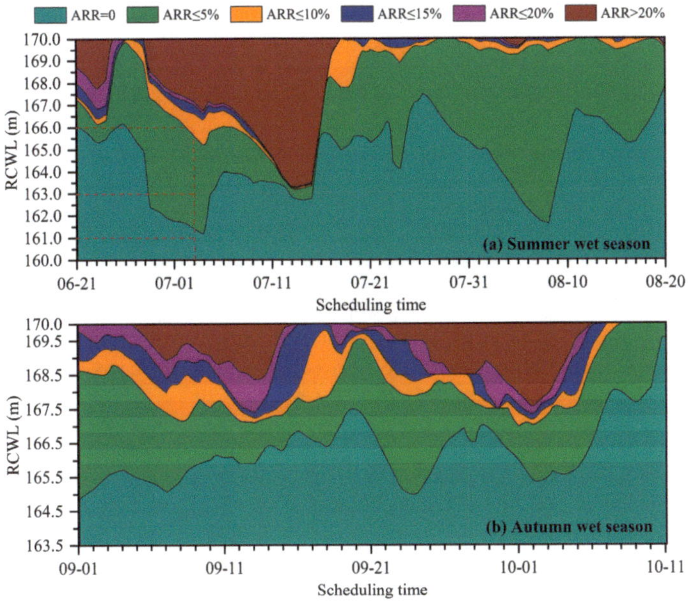


**Figure 8.** Flood risk control maps of the Danjiangkou reservoir. (**a**) Summer wet season; (**b**) autumn wet season.

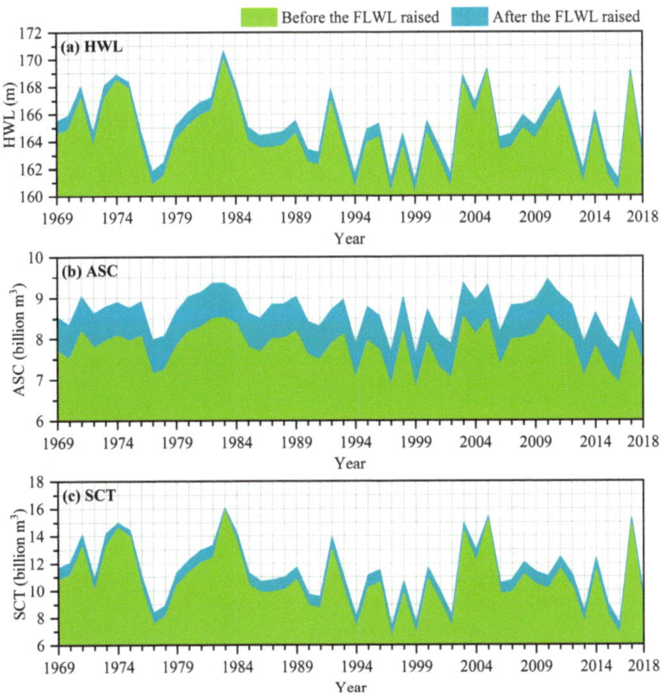


**Figure 9.** The changes in the HWL, ASC, and SCT values before and after the FLWL was raised. (**a**) The HWL values; (**b**) the ASC values; (**c**) the SCT values.

Figure 9b demonstrates that the ASC values all increased after the FLWL was raised. The annual mean and maximum of the ASC are 8.64 and 9.43 billion m$^3$, respectively, and the ASC values are, on average, 10.68% (9.66–12.13%) more than the ones from before the FLWL was raised. In addition, the ASC values are, on average, 0.83 billion m$^3$ more than the ones before the FLWL was raised, which is 8.74% of the average annual volume (9.5 billion m$^3$) of water transferred by the aforementioned water transfer project over many years. As a consequence, the average available flood resources during the wet season increased after the FLWL was raised.

Figure 9c illustrates that the SCT values are all higher than the ones from before the FLWL was raised. The annual mean and maximum of the SCT are 11.40 and 16.01 billion m$^3$, respectively, and the SCT values are, on average, 8.13% (1.10–12.78%) more than the ones from before the FLWL was raised. Moreover, the SCT are values, on average, 0.81 billion m$^3$ more than the ones from before the FLWL was raised, which is 8.53% of the average annual volume of water transferred. Hence, the average available flood resources at the end of the wet season increased after the FLWL was raised.

## 5. Discussion

During reservoir operation and management, the flood risk control method oriented towards floodwater utilization can be directly applied to the DC-FLWL on the basis of the flood risk control map (Figure 8). Assuming an acceptable risk ratio of 5%, taking the summer wet season as an example, if the water level on 3 July is 166.00 m, the risk ratio exceeds 5% but is not higher than 10% (as indicated by the auxiliary line in Figure 8a), and the water level could be decreased adequately in the green area to reduce flood risk. Conversely, if the water level on 3 July is 161.00 m, the risk ratio does not exceed the acceptable risk ratio of 5%, and the water level may be prompted to the green area to enhance floodwater utilization based on the forecast information. The aim of this flood risk control method is to realize the DC-FLWL under an acceptable risk ratio, thereby enhancing floodwater utilization as much as possible while ensuring an acceptable risk level.

Furthermore, the risk ratios of various FLWLs were studied for the creation of this paper to verify the reasonability of the dynamic control domains of the FLWL under various acceptable risk ratios. The FLWL was raised in increments of 0.50 m, and the risk ratios of various FLWLs were calculated using a risk analysis model on the basis of 30,000 simulated floods. Figure 10 demonstrates more detailed information about the changes in risk ratio under various FLWLs and flood return periods (5a, 10a, 20a, 50a, 100a).

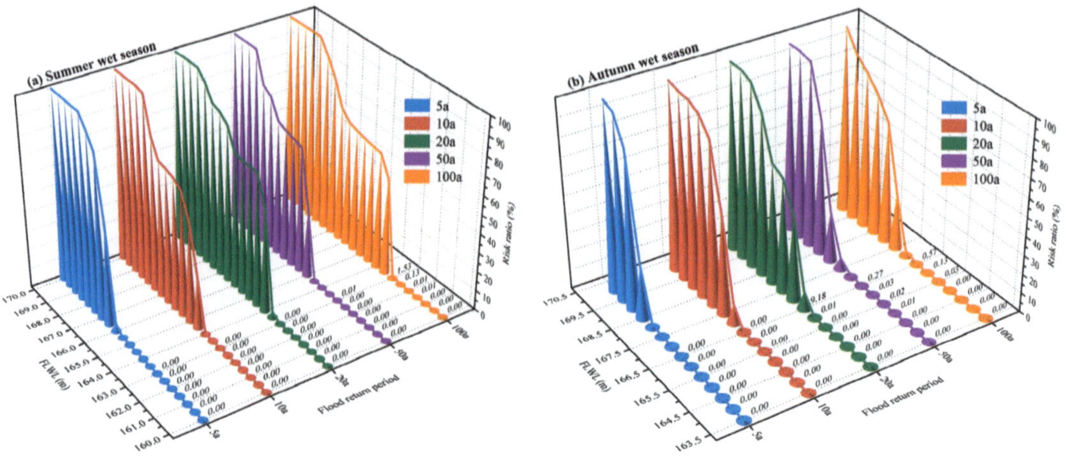


**Figure 10.** The changes in risk ratio under various FLWLs and flood return periods. (**a**) Summer wet season; (**b**) Autumn wet season.

The results in Figure 10 reveal that the risk ratio gradually increases as the FLWL increases, and Figure 9 also indicates that the both risks and benefits increased as FLWL rose, which is in line with some existing studies [33–35]. Specifically, taking the flood of shown by the red curve and red cones in Figure 10a for the summer wet season as an example, the risk ratio remains relatively constant when the FLWL is less than 164.00 m. However, the risk ratio steeply increases with increasing FLWL when the FLWL is greater than 164.00 m. It is important to note that the risk ratios are 0 when the FLWLs are not more than 161.00 m and 164.50 m, respectively, during the summer and autumn wet seasons, which validates the potential of raising the FLWL in another way. This means that when the FLWL is raised from the current FLWL by 1.00 m, the flood risks for floods that occur once in 100 years and less remain unchanged and are all 0, which is the reason why many studies aim to enhance floodwater utilization without increasing the flood risk [36,37]. Whilst raising the FLWL has the potential to increase flood risk, this rule is not absolute. The flood risk may remain the same after raising the FLWL in some cases [38–40].

From what has been discussed above, it can be concluded that the flood risk is within a controllable range when the FLWL is raised by 1.00 m, based on the present study's FLWL, and the ASC and SCT both increased (annual average increase of 0.83 and 0.81 billion $m^3$, respectively), according to Figure 9, which effectively improves floodwater utilization and contributes to guaranteeing the water supply and power generation of the Danjiangkou reservoir.

## 6. Conclusions

Realizing the DC-FLWL is one of the ways of enhancing floodwater utilization, and the focus of this approach is to balance flood risk and the benefits of floodwater utilization. This paper presents a flood risk control method oriented towards floodwater utilization that achieves the dynamic control domains of the FLWL under various acceptable risk ratios, considering multiple main risk factors. Based on the comprehensive attention paid in this paper to the main flood risk factors, including the uncertainty of reservoir inflow, interval floods, and flood forecast errors, a risk analysis model was constructed, after which the process of calculating the RCWL was proposed to acquire the reservoir flood risk control map, both which were used in the flood risk control method oriented towards floodwater utilization. Then, the risks and benefits of floodwater utilization were investigated. Finally, the Danjiangkou reservoir was used as a case study. Based on our study, the following conclusions can be drawn.

(1) The dynamic control domains of the FLWL under various acceptable risk ratios and flood risk control maps were obtained, and the flood risk control method could improve floodwater utilization by realizing the DC-FLWL under acceptable risk ratios.
(2) The potential of raising the FLWL was studied by using the established risk analysis model. The results show that the risk ratio increases as the FLWL rises. The FLWL could be raised by 1.00 m based on the present FLWL during the summer and autumn wet seasons under the absence of flood risk.
(3) The assessment criteria, namely the HWL, ASC, and SCT values, were used to quantitatively measure the risks and benefits of floodwater utilization. The flood risk is within an acceptable range, and the available flood resources in the wet and dry seasons could be increased by about 10.68% and 8.13%, respectively, which could effectively improve the floodwater utilization and contribute to water supply safety.

The flood risk control method proposed in this paper achieves the dynamic control of the FLWL to improve floodwater utilization under the acceptable flood risk. However, amidst the backdrop of global climate change, considering key risk factors as comprehensively as possible, achieving the DC-FLWL, and weighing the risks and benefits of floodwater utilization remain as challenges for the future.

**Author Contributions:** Conceptualization, L.Z. and L.K.; data curation, L.Z.; formal analysis, L.Z.; funding acquisition, L.K.; investigation, S.H. and J.G.; methodology, L.Z.; project administration,

L.K.; resources, L.K.; software, L.Z., S.H. and J.G.; supervision, L.K.; validation, L.K., S.H. and J.G.; visualization, L.Z.; writing—original draft, L.Z.; writing—review and editing, L.K. All authors have read and agreed to the published version of the manuscript.

**Funding:** This paper was funded by the National Key R&D Program of China (Grant No. 2021YFC3200302).

**Data Availability Statement:** The data that support the findings of this paper are available from the corresponding author upon reasonable request.

**Acknowledgments:** We would like to express our gratitude to the Hanjiang Water Conservancy and Hydropower (Group) Co., Ltd. (Danjiangkou, China) for providing data.

**Conflicts of Interest:** The authors declare no conflict of interest.

## References

1. Garrote, J.; Pena, E.; Diez-Herrero, A. Probabilistic Flood Hazard Maps from Monte Carlo Derived Peak Flow Values—An Application to Flood Risk Management in Zamora City (Spain). *Appl. Sci.* **2021**, *11*, 6629. [CrossRef]
2. Zhou, L.; Kang, L. A Comparative Analysis of Multiple Machine Learning Methods for Flood Routing in the Yangtze River. *Water* **2023**, *15*, 1556. [CrossRef]
3. Wen, Y.; Zhou, L.; Kang, L.; Chen, H.; Guo, J. Drought risk analysis based on multivariate copula function in Henan Province, China. *Geomat. Nat. Hazards Risk* **2023**, *14*, 2223344. [CrossRef]
4. Fadel, A.W.; Marques, G.F.; Goldenfum, J.A.; Medellín-Azuara, J.; Tilmant, A. Full Flood Cost: Insights from a Risk Analysis Perspective. *J. Environ. Eng.* **2018**, *144*, 04018071. [CrossRef]
5. Ding, W.; Zhang, C.; Peng, Y.; Zeng, R.; Zhou, H.; Cai, X. An analytical framework for flood water conservation considering forecast uncertainty and acceptable risk. *Water Resour. Res.* **2015**, *51*, 4702–4726. [CrossRef]
6. Li, Y.; Guo, S.L.; Quo, J.L.; Wang, Y.; Li, T.Y.; Chen, J.H. Deriving the optimal refill rule for multi-purpose reservoir considering flood control risk. *J. Hydro-Environ. Res.* **2014**, *8*, 248–259. [CrossRef]
7. Liu, G.; Zhou, J.; Yang, Y.; Ke, S.; Mo, L.; Wang, Y.; Xiao, X. Risks Analysis of Discarding Water in Power Generation Process of Hydropower Station. *IEEE Access* **2020**, *8*, 132703–132720. [CrossRef]
8. Zhang, J.; Min, Y.W.; Feng, B.F.; Duan, W.X. Research and Application of Key Technologies for Dynamic Control of Reservoir Water Level in Flood Season. *Water* **2021**, *13*, 3576. [CrossRef]
9. Ding, W.; Zhang, C.; Cai, X.M.; Li, Y.; Zhou, H.C. Multiobjective hedging rules for flood water conservation. *Water Resour. Res.* **2017**, *53*, 1963–1981. [CrossRef]
10. Tan, Q.F.; Wang, X.; Liu, P.; Lei, X.H.; Cai, S.Y.; Wang, H.; Ji, Y. The Dynamic Control Bound of Flood Limited Water Level Considering Capacity Compensation Regulation and Flood Spatial Pattern Uncertainty. *Water Resour. Manag.* **2017**, *31*, 143–158. [CrossRef]
11. Zhou, Y.L.; Guo, S.L.; Chang, F.J.; Liu, P.; Chen, A.B. Methodology that improves water utilization and hydropower generation without increasing flood risk in mega cascade reservoirs. *Energy* **2018**, *143*, 785–796. [CrossRef]
12. Zhang, Z.B.; He, X.Y.; Geng, S.M.; Zhang, S.H.; Ding, L.Q.; Kan, G.Y.; Li, H.; Jiang, X.M. An Improved "Dynamic Control Operation Module" for Cascade Reservoirs. *Water Resour. Manag.* **2018**, *32*, 449–464. [CrossRef]
13. Gong, Y.; Liu, P.; Cheng, L.; Chen, G.Y.; Zhou, Y.L.; Zhang, X.Q.; Xu, W.F. Determining dynamic water level control boundaries for a multi-reservoir system during flood seasons with considering channel storage. *J. Flood Risk Manag.* **2020**, *13*, 17. [CrossRef]
14. Ning, Y.W.; Ding, W.; Liang, G.H.; He, B.; Zhou, H.C. An Analytical Risk Analysis Method for Reservoir Flood Control Operation Considering Forecast Information. *Water Resour. Manag.* **2021**, *35*, 2079–2099. [CrossRef]
15. Mu, Z.; Ai, X.; Ding, J.; Huang, K.; Chen, S.; Guo, J.; Dong, Z. Risk Analysis of Dynamic Water Level Setting of Reservoir in Flood Season Based on Multi-index. *Water Resour. Manag.* **2022**, *36*, 3067–3086. [CrossRef]
16. Pan, J.B.; Xie, Y.; Liu, M.; Gao, Z.; Gao, Z.Z.; Xue, B.W. Dynamic Control of Water Level in Flood Limited Reservoir Based on Intelligent Calculation. *Math. Probl. Eng.* **2022**, *2022*, 10. [CrossRef]
17. Lu, Q.; Zhong, P.; Xu, B.; Huang, X.; Zhu, F.; Wang, H.; Ma, Y. Multi-objective risk analysis for flood control operation of a complex reservoir system under multiple time-space correlated uncertainties. *J. Hydrol.* **2022**, *606*, 127419. [CrossRef]
18. Du, H.; Wang, Z.; Yin, J. Analysis of flood control risk in floodwater utilization considering the uncertainty of flood volume and peak. *Front. Earth Sci.* **2023**, *11*, 1196903. [CrossRef]
19. Chen, J.; Zhong, P.A.; Zhang, W.G.; Zhu, F.L.; Zhang, Y. Improved Risk-Assessment Model for Real-Time Reservoir Flood-Control Operation. *J. Water Resour. Plan. Manag.* **2020**, *146*, 0001174. [CrossRef]
20. Winter, B.; Schneeberger, K.; Huttenlau, M.; Stotter, J. Sources of uncertainty in a probabilistic flood risk model. *Nat. Hazards* **2018**, *91*, 431–446. [CrossRef]
21. Peng, Y.; Chen, K.; Yan, H.; Yu, X. Improving Flood-Risk Analysis for Confluence Flooding Control Downstream Using Copula Monte Carlo Method. *J. Hydrol. Eng.* **2017**, *22*, 04017018. [CrossRef]
22. Wang, Q.S.; Zhou, J.Z.; Dai, L.; Huang, K.D.; Zha, G. Risk assessment of multireservoir joint flood control system under multiple uncertainties. *J. Flood Risk Manag.* **2021**, *18*, e12740. [CrossRef]

23. Barría, P.; Chadwick, C.; Ocampo-Melgar, A.; Galleguillos, M.; Garreaud, R.; Díaz-Vasconcellos, R.; Poblete, D.; Rubio-Álvarez, E.; Poblete-Caballero, D. Water management or megadrought: What caused the Chilean Aculeo Lake drying? *Reg. Environ. Change* **2021**, *21*, 19. [CrossRef]
24. Valencia, D.; Schakke, J.C. Disaggregation processes in stochastic hydrology. *Water Resour. Res.* **1973**, *9*, 580–585.
25. Chen, J.; Zhong, P.; Zhao, Y.; Xu, B. Risk analysis for the downstream control section in the real-time flood control operation of a reservoir. *Stoch. Environ. Res. Risk Assess.* **2015**, *29*, 1303–1315. [CrossRef]
26. Fan, Y.R.; Huang, W.W.; Huang, G.H.; Li, Y.P.; Huang, K.; Li, Z. Hydrologic risk analysis in the Yangtze River basin through coupling Gaussian mixtures into copulas. *Adv. Water Resour.* **2016**, *88*, 170–185. [CrossRef]
27. Kang, L.; Jiang, S. Bivariate Frequency Analysis of Hydrological Drought Using a Nonstationary Standardized Streamflow Index in the Yangtze River. *J. Hydrol. Eng.* **2019**, *24*, 05018031. [CrossRef]
28. Li, W.; Zhou, J.; Yao, X.; Feng, K.; Luo, C.; Sun, N. Flood Hazard Analysis Based on Copula Connect Function. *Nat. Hazards Rev.* **2023**, *24*, 04022041. [CrossRef]
29. Lu, Q.; Zhong, P.-a.; Xu, B.; Zhu, F.; Ma, Y.; Wang, H.; Xu, S. Risk analysis for reservoir flood control operation considering two-dimensional uncertainties based on Bayesian network. *J. Hydrol.* **2020**, *589*, 125353. [CrossRef]
30. Dai, L.; Zhou, J.; Chen, L.; Huang, K.; Wang, Q.; Zha, G. Flood-risk analysis based on a stochastic differential equation method. *J. Flood Risk Manag.* **2019**, *12*, e12515. [CrossRef]
31. Chen, G.; Lin, H.; Hu, H.; Yan, Y.; Wan, Y.; Xiao, T.; Peng, Y. Research on the Measurement of Ship's Tank Capacity Based on the Monte Carlo Method. *Chem. Technol. Fuels Oils* **2022**, *58*, 232–236. [CrossRef]
32. Kang, L.; He, X. Risk analysis of rich–poor rainfall encounter in inter-basin water transfer projects based on Bayesian networks. *Water Supply* **2011**, *11*, 451–459. [CrossRef]
33. Chang, J.; Guo, A.; Du, H.; Wang, Y. Floodwater utilization for cascade reservoirs based on dynamic control of seasonal flood control limit levels. *Environ. Earth Sci.* **2017**, *76*, 260. [CrossRef]
34. Yang, Z.; Huang, X.; Liu, J.; Fang, G. Optimal Operation of Floodwater Resources Utilization of Lakes in South-to-North Water Transfer Eastern Route Project. *Sustainability* **2021**, *13*, 4857. [CrossRef]
35. Wang, K.; Wang, Z.; Xu, J.; Mo, Y.; Cheng, L.; Bai, Y.; Du, H.; Liu, K. Floodwater utilization potential assessment of China based on improved conceptual model and multi-reservoir basin assessment method. *J. Hydrol.* **2023**, *625*, 129921. [CrossRef]
36. Li, X.; Guo, S.; Liu, P.; Chen, G. Dynamic control of flood limited water level for reservoir operation by considering inflow uncertainty. *J. Hydrol.* **2010**, *391*, 124–132. [CrossRef]
37. Ye, A.; Wang, Z.; Zhang, L.; Wang, L.; Wang, K. Assessment approach to the floodwater utilization potential of a basin and an empirical analysis from China. *Environ. Earth Sci.* **2019**, *78*, 52. [CrossRef]
38. Xie, A.; Liu, P.; Guo, S.; Zhang, X.; Jiang, H.; Yang, G. Optimal Design of Seasonal Flood Limited Water Levels by Jointing Operation of the Reservoir and Floodplains. *Water Resour. Manag.* **2018**, *32*, 179–193. [CrossRef]
39. Lu, Q.; Zhong, P.-a.; Xu, B.; Zhu, F.; Huang, X.; Wang, H.; Ma, Y. Stochastic programming for floodwater utilization of a complex multi-reservoir system considering risk constraints. *J. Hydrol.* **2021**, *599*, 126388. [CrossRef]
40. Wang, K.; Wang, Z.; Liu, K.; Du, H.; Mo, Y.; Wang, M.; Jin, G.; Li, L. Assessing the floodwater utilization potential in a reservoir-controlled watershed: A novel method considering engineering regulations and an empirical case from China. *Ecol. Inform.* **2022**, *68*, 101581. [CrossRef]

water

MDPI

Article

# Evolutionary Trends and Coordinated Development Analysis of Water Resources Systems and High-Quality Economic Growth in the Yangtze River Delta

Di Liu [1,†], Qin Dai [2,*,†] and Guanghui Yuan [3,†]

1 Institute of Finance and Economics/School of Urban and Regional Sciences, Shanghai University of Finance and Economics, Shanghai 200433, China; ldi2301@126.com
2 School of Economics and Management, Shanghai University of Electric Power, Shanghai 200090, China
3 School of Economics and Management, Shanghai University of Political Science and Law, Shanghai 201701, China; guanghuiyuan@outlook.com
* Correspondence: qindai@shiep.edu.cn
† These authors contributed equally to this work.

**Abstract:** This article calculates the indices for high-quality economic development and water resource systems across 25 cities in the Yangtze River Delta from 2011 to 2021. Utilizing a multifaceted analytical framework comprising the CRITIC method, standard deviation ellipse, harmonious development coefficient, and coupling coordination coefficient, we investigate spatiotemporal evolutionary trends and overarching harmonious development states between the two systems. Results indicate: (1) Throughout the research period, mean values of high-quality economic development indices fluctuated within the range of 0.05 to 0.68, while water resource carrying capacity indices oscillated between 0.18 and 0.81. (2) The epicenter of high-quality economic development indices is situated in the periphery of Lake Tai, whereas the fulcrum of the water resource system indices is located in Huzhou City, both displaying a northwest-southeast orientation. (3) Coupling coordination development exhibits a propitious advancement trajectory, with certain locales attaining exemplary coordinated growth.

**Keywords:** high-quality economic development; water resource systems; standard deviation ellipse; coupling coordination development

## 1. Introduction

The issue of high-quality economic development and the developmental trends in water resource carrying capacity is a comprehensive issue, which is included in the process of researching the sustainable development path of society–economy–resources–environment. Amidst the acceleration towards an epoch characterized by high-quality economic expansion, harmonizing the relationship between water resource utilization efficacy and superior economic growth becomes an exigent quandary warranting immediate redress. The 2018 Yangtze River Delta Leadership Roundtable emphasized "Focusing on Quality, Fostering Integration" as its thematic cornerstone, delineating the blueprint for integrated high-quality development in the region. Serving as a pivotal strategic overlay within the Delta, the Yangtze River Delta is primed not only to advance its own elevated economic development but also to actuate industrial transference and diffusion to the surrounding area, thereby engendering high-quality growth in neighboring provinces. The National Development and Reform Commission, in its "Socioeconomic Development Report" in 2017 and 2018, explicitly underscores the imperative for high-quality development to engender a resource-efficient and ecologically benign development architecture. To carve out a focal zone of exceptional economic development within the Yangtze River Delta, there is a compelling necessity to pivot away from an unsustainable growth model predicated upon high input, consumption, and pollution, towards augmenting the efficiency of input factors. Policy documents such as the "Yangtze River Delta Regional Integration Development Plan" and the "Yangtze River Delta Ecological Green Integrated Development Demonstration Zone Overall Plan" advocate the pre-emptive transmutation of ecological assets into socio-economic developmental advantages. Currently, the water demand in the Yangtze River Delta region is increasing, while the increasingly severe water shortage and water environment degradation problems are seriously constraining the sustainable development of the regional industrial economy. At the same time, the uneven allocation of water resources in the region leads to the low economic value of water resources, and the utilization of water resource flow capital and stock capital, as the core of the three-dimensional water resource ecological footprint model, are closely related to economic development. The Yangtze River Delta, as the first urban agglomeration in China, is a strategic area for China's economic development and also a sensitive area for environmental protection, and the coordination between its economy and environment is particularly important, and it is significant to explore the coordination relationship between the economy and the environment based on the perspective of a large watershed in the evolution of spatial and temporal patterns.

With the advancement of urbanization, the traditional provincial and administrative economies are changing to urban agglomerations, which are important for promoting regional economic development and water resource management. Therefore, it is urgent to analyze the correlation between water resources and economic development at the scale of urban agglomerations, so as to reveal the efficiency of water resource use and the impact of economic development on the sustainable use of water resources. The correlation between water resources and economic development at the scale of urban agglomerations is therefore urgent. Therefore, our manuscript concurrently calculates indices for high-quality economic development and water resource systems, employing coupling coefficients and standard deviation ellipses to gauge the spatio-temporal distribution and centroidal dispersion of high-quality economic development and water resource systems, respectively. It delves into their spatial evolutionary trends within the period under investigation and calculates their degree of coordinated development (coupled coordination and elliptic difference models). The assessment is undertaken from both spatiotemporal matching trends and quantitative synergy perspectives, aiming to elucidate the operative mechanisms under various states of coordination. This serves as a foundational reference for enhancing water resource utilization efficiency, expediting high-quality economic advancement, and facilitating the amelioration of urban aquatic environments, thereby mitigating the inherent contradictions within the water resource high-quality economic development system to achieve sustainable progress in both domains.

## 2. Literature Overview

Presently, scholars both domestically and internationally have embarked on extensive research endeavors to examine the intricate interrelationship between water resources and economic development, yielding a myriad of significant findings. International research concerning the nexus between economy and environment commenced in the mid-20th century [1–3], concentrating primarily on an array of focal points such as the societal and economic ramifications on water resource demand and impact, and the role and limitations of water resources in the progression of urbanization, as well as the implications of water resources provision on economic proliferation [1–8]. Ren et al. [9] established a system dynamics model of water resources, water environment, and water ecological carrying capacity, and explored the relationship between water resources and ecology in the Wulansu Sea by simulating five scenarios. Luo et al. [10] constructed an economic–water–ecological framework and established a harmonious regulation model to explore the degree of harmonious development in the Shaying River Basin. The research about China in this realm originated in the 1970s, with environmental economics serving as the theoretical cornerstone. A large number of studies [11–14] have led to a rich discussion around the existence of environmental Kuznets curves. These studies have pointed out that achieving harmonious and sustainable development of the economy and the environment is key to the social stability of countries. Given the complexities of demographic structure, economic systems, and disparate resource allocation, Chinese scholars [15–17] have undertaken rigorous analyses of water resources and economic growth. Situating urban development as the backdrop for research, multifarious methodologies such as Fuzzy Set Theory, Multi-objective Models of Aquatic Ecological Carrying Capacity, and Semi-Γ Distribution Indices based on Particle Swarm Optimization have been employed to scrutinize varying scales, including urban water bodies, metropolitan clusters, catchment areas, and provincial domains, in locations such as the Henan Province, East Lake in Wuhan, Changsha-Zhuzhou-Xiangtan Urban Agglomeration, Yangtze River Delta Catchment, and Jiangsu Province. These studies [18–20] aim to evaluate the water environment's carrying capacity, the responsiveness of urban clusters to water resources, water level fluctuations, and the carrying capacity of urban water resources, in an endeavor to elucidate sustainable models for water resource management. Furthermore, treating economic development and water resources as discrete yet interrelated systems for coupled research constitutes another investigative modality for deciphering their mutual interplay [21,22]. Zhang et al. [23] used the four major river basins in the Henan Province as the research scale, and analyzed the evolution of the coupled coordination degree of water resources use and economic and social development by using the coupled coordination degree model and gray prediction method.

In summary, research focusing individually on water resources and economic development is relatively prevalent, encompassing theoretical frameworks, evaluative assessments of developmental phases, and case studies. The methodologies employed range from conventional trend analysis to computationally intensive techniques based on the Environmental System of Equations Framework (ESEF), system dynamics models, and neural network-based models of water resource carrying capacity [24–26]. However, current studies that consider both elements predominantly scrutinize the development of one aspect through the lens of the other, and there is a notable lack of comprehensive exploration of the mechanisms of their interaction, synergistic growth, and influencing factors. Thus, it becomes challenging to holistically appraise their regional compatibility and specific impact mechanisms. Our marginal contribution lies in the fact that the study is based on the Yangtze River Delta urban agglomeration, which has the most developed water resource system and economic development system, and the study of the coupling of resources and economy in this region can help other regions and urban agglomerations in China to explore new development paths. Secondly, the CRITIC methodology is used to measure the high-quality development of the water resource system and economy, to rationally judge the current development status of each city from the perspective of objective empowerment, and to help the government to formulate the direction of local green development.

## 3. Overview of the Research Region

The Yangtze River Delta, situated at the downstream confluence of the Yangtze River, flanked by the Yellow Sea and the East China Sea, constitutes an ecotone characterized by a maritime–terrestrial interface. Comprising two provinces, Jiangsu and Zhejiang, along with the municipality of Shanghai, the region spans an expansive geographical expanse of approximately 219,000 square kilometers. The climatic conditions predominantly exhibit subtropical monsoonal characteristics, with elevated warming rates primarily observed during the winter and spring months, and a relative attenuation during the summer. Trends in temperature fluctuations, warming rates, and the contributions to warming from urbanization effects exhibit a commendable congruence with those of other regions. Hydrologically, the Yangtze River Delta boasts an intricate network of water bodies, featuring prominent lakes such as Taihu, Hongze, and Jinniu in Jiangsu, and Xihu, Donghu, and Qiandao in Zhejiang. In addition to significant rivers like the Yangtze, Qiantang, and the Grand Canal, the area is also replete with other noteworthy waterways such as Shanghai's Huangpu and Wusong rivers, Jiangsu's Qinhuai and Xinshu rivers, and Zhejiang's Oujiang and Lingjiang rivers. The region is further enriched by a plethora of lacustrine resources, riparian zones, and wetlands. These aquatic resources play an indispensable role in sustaining agricultural irrigation and industrial development within the Yangtze River Delta milieu.

The Yangtze River Delta, centered on Shanghai and radiating to the surrounding cities, has become one of the most dynamic and economically developed regions in the world's major urban agglomerations. In 2022, Jiangsu's GDP exceeded the 12 trillion-yuan mark and its economic scale stepped up to a new level, while Zhejiang's GDP exceeded 7.5 trillion yuan and Shanghai's GDP exceeded 4 trillion yuan. Based on the country's economic policies and as a region created by the state to become rich first, the Yangtze River Delta has achieved remarkable economic results and provided a good model for China's economic development. In recent years, under the guidance of the country's new development philosophy, the economic development of the Yangtze River Delta region has also turned to green, environmental protection, and sustainability.

## 4. Indicator Framework and Research Methodology

*Indicator Framework*

In alignment with the unique attributes of the Yangtze River Delta, we provisionally select an evaluative index system to gauge the levels of high-quality economic development and water resource systems. Utilizing the CRITIC method, we ascertain the weights of these indices. After determining the weights, we labeled the attributes in Table 1 with reference to existing studies and the relevance of each indicator. The "+" represents that the indicator layer will have a positive impact on the target layer and the "−" represents that the indicator layer will have a negative impact on the target layer. The water resource system encompasses three dimensions, namely water resource availability, utilization, and conservation, and includes a total of seven indicators. On the other hand, high-quality economic development is segmented into four dimensions: industrial structure, technological innovation, ecological environment, and standard of living, comprising a total of nine indicators. Within this, the index of industrial advancement is calculated based on the ratio of the output value of tertiary to secondary sectors. The rationalization of the industrial structure is assessed by drawing upon methodologies proposed by Deng Huihui et al. [27]. Given the data paucity regarding the output value of the productive service sector in most cities, the proportion of employment in productive service industries (productive services industries mainly include R&D, design and other technical services for production activities, cargo transportation, general aviation production, warehousing and postal courier services, information services, financial services, energy-saving and environmental protection services, productive leasing services, business services, human resource management and vocational education and training services, wholesale and trade brokerage and agency services, and productive support services) is adopted as a surrogate metric.

**Table 1.** Indicator system for evaluating water resource system and high-quality economic development.

| Objective Layer | Criteria Layer | Indicator Layer | Unit | Weight | Attribute |
|---|---|---|---|---|---|
| Water Resource System | Water Resource Availability | Total Water Resources | 10,000 $m^3$ | 0.150 | + |
| | | Annual Precipitation | 10,000 $m^3$ | 0.170 | + |
| | Water Resource Utilization | Total Urban Water Supply | 10,000 $m^3$ | 0.169 | + |
| | | Domestic Water Consumption per Capita | 10,000 $m^3$ | 0.152 | + |
| | | Industrial Water Consumption | 10,000 $m^3$ | 0.102 | − |
| | Water Resource Conservation | Industrial Wastewater Discharge | 10,000 $m^3$ | 0.057 | − |
| | | Wastewater Treatment Rate | % | 0.200 | + |

**Table 1.** *Cont.*

| Objective Layer | Criteria Layer | Indicator Layer | Unit | Weight | Attribute |
|---|---|---|---|---|---|
| High-Quality Economic Development | Industrial Structure | Industrial Advancement Index | - | 0.104 | + |
| | | Rationalization of Industrial Structure | - | 0.110 | + |
| | | Proportion of Productive Service Sector | % | 0.132 | + |
| | Technological Innovation | Number of Patents Granted | Individual | 0.097 | + |
| | Ecological Environment | Green Coverage Rate in Built-up Areas | % | 0.103 | + |
| | | Comprehensive Utilization Rate of Industrial Solid Waste | % | 0.103 | + |
| | Standard of Living | Per Capita GDP | Yuan/Person | 0.155 | + |
| | | Per Capita Education Expenditure | Yuan/Person | 0.108 | + |
| | | Hospital Bed Availability per 10,000 People | Beds/10,000 People | 0.089 | + |

Note: Source: Statistical Yearbook and Statistical Bulletin of Municipalities.

## 5. Research Methodology

### 5.1. Refinement of the CRITIC Method

In the selection of water resource system evaluation models and methods, representative research methods mainly include subjective empowerment methods such as principal component analysis [28], the fuzzy comprehensive evaluation method [29] and objective empowerment methods such as the BP neural network [30], data envelopment analysis method [31] and Topsis [32]. Subjective empowerment is subject to human influence, so we choose the objective empowerment method. The CRITIC (Criteria Importance Through Inter-criteria Correlation) methodology, originally postulated by Diakoulaki (1995), serves as an objective algorithm for attribute weighting. This method harnesses the differential attributes and inter-criterion discordances to reflect the informational content and distinctiveness of each criterion, consequently establishing their respective weights. Variability is for the differences in the magnitude of the values of the same indicator among different samples, which is determined by the standard deviation of each indicator, and the standard deviation is calculated for each column after preprocessing. The size and direction of conflict is expressed by the correlation coefficient, which can be expressed as $\sum_{i=1}^{n}(1-|r_{ij}|)$, where $r_{ij}$ is the correlation coefficient between the jth indicator and the ith indicator, and it also indicates that for the positive correlation and the negative correlation which have the same absolute value, the conflict between the indicators is the same. If two indicators have a strong positive correlation, it means that their conflictability is low. The weight of the indicator is calculated according to the difference and conflict of the indicator, and let $C_j$ be the amount of information contained in the jth indicator, and the weight of indicator j is the proportion of the information $C_j$ (i.e., the product of the difference and conflict) contained in indicator j to the proportion of all of the information, which is expressed as follows:

$$C_j = \frac{\sigma_j}{\overline{x}}\sum_{i=1}^{n}(1-|r_{ij}|) \quad j = 1, 2, \ldots, n \tag{1}$$

where a larger $C_j$ signifies a higher informational content and correspondingly a greater weight for the j criterion.

$$W_j = \frac{C_j}{\sum_{i=1}^{n}(C_j)} \quad j = 1,2,\ldots,n \tag{2}$$

### 5.2. Coupling Coordination Model

The Coupling Coordination Model delineates the degree of mutual influence and synergistic advancement between two or more interconnected systems, thereby capturing the equilibrium state of water resource utilization in concert with high-quality economic development. The computational formula is as follows.

$$C = \sqrt{\frac{f(x) \times f(y)}{\left[\frac{f(x)+f(y)}{2}\right]^2}} \tag{3}$$

$$T = \alpha f(x) + \beta f(y) \tag{4}$$

$$D = \sqrt{C \times T}$$

where $D$ represents the coupling coordination degree, C signifies the coupling correlation degree, and T denotes the comprehensive evaluation index. $f(x)$ and $f(y)$ are evaluative values for high-quality economic development and water resource systems, respectively. The variables $\alpha$ and $\beta$ manifest the relative contributions of high-quality economic development and water resource utilization. We also used the CRITIC method to assign weights and calculate the standard deviation and correlation of the indicators, obtaining $\alpha = 0.484$ and $\beta = 0.516$. Drawing upon extant research findings [33], the phase and type of synergistic development between water resource utilization and high-quality economic development in the Yangtze River Delta are articulated in Table 2.

**Table 2.** Classification criteria for coupling coordination degrees.

| Range of Coupling Coordination Degree D Value | Level of Coupling Coordination | Developmental Tier | Coordinated Developmental Tier |
|---|---|---|---|
| [0.0~0.3] | Severe Disequilibrium | Severe Regression | Critical Disequilibrium and Regression |
| (0.3~0.4] | Moderate Disequilibrium | Moderate Regression | Moderate Disequilibrium and Regression |
| (0.4~0.5] | Mild Disequilibrium | Mild Regression | Mild Disequilibrium and Regression |
| (0.5~0.55] | Moderate Coordination | Moderate Development | Moderate Coordinated Advancement |
| (0.55~0.7] | Good Coordination | Favorable Development | Favorably Coordinated Advancement |
| (0.7~1.0] | Optimal Coordination | Optimal Development | Optimal Coordinated Advancement |

### 5.3. Relative Development Model

While the Coupling Coordination Model elucidates the degree of synchronized advancement between high-quality economic growth and water resources, it falls short of quantifying the developmental disparity between the two entities. Consequently, the relative development index is introduced to delineate the level of either accelerated or lagging development between them. The mathematical representation for the relative development index is as follows.

$$P = \frac{f(x)}{f(y)} \tag{5}$$

In this equation, $P$ denotes the relative development index. When $P > 1.2$, it signifies that the regional high-quality economic growth is in a state of accelerated development. If $0.8 < P \leq 1.2$, it indicates that high-quality economic growth and water resources are advancing in synchrony. Should $P \leq 0.8$, it portrays the regional high-quality economic growth as lagging.

### 5.4. Standard Deviation Ellipse Model

The standard deviation ellipse model falls under the purview of spatial pattern statistical analysis, focusing predominantly on explicating the global characteristics of geographical feature distributions. The precise computational formulae for the parameters of the standard deviation ellipse are delineated as follows.

$$x' = x_i - x_{ave}; y' = y_i - y_{ave} \tag{6}$$

$$tan\theta = \frac{\left(\sum_{i=1}^{n} W_i^2 x_i'^2 - \sum_{i=1}^{n} W_i^2 y_i'^2\right) + \sqrt{\sum_{i=1}^{n} W_i^2 x_i' y_i' - \sum_{i=1}^{n} W_i^2 y_i'^2 + 4\left(\sum_{i=1}^{n} W_i^2 x_i' y_i'\right)}}{2\sum_{i=1}^{n} W_i^2 x_i^2 y_i^2} \tag{7}$$

$$\delta_x = \sqrt{\frac{\sum_{i=1}^{n} \left(W_i x_i' \cos\theta - W_i y_i' \sin\theta\right)^2}{\sum_{i=1}^{n} W_i'^2}} \tag{8}$$

$$\delta_y = \sqrt{\frac{\sum_{i=1}^{n} \left(W_i x_i' sin\theta - W_i y_i' cos\theta\right)^2}{\sum_{i=1}^{n} W_i^2}} \tag{9}$$

where $(x_{ave}, y_{ave})$ represents the centroid of the coordinates $(x_i, y_i)$; $W_i$ denotes the indices for high-quality regional economic development as well as water resource systems; $x'$ and $y'$ symbolize the relative coordinates of individual points about the regional centroid. The angle of rotation for the centroidal distribution pattern can be ascertained based on $tan\theta$; and $\delta_x$ and $\delta_y$ are the standard deviations along the X-axis and Y-axis, respectively.

### 5.5. Data Sources

The hydrological data principally originate from the water resource bulletins and statistical yearbooks of various municipalities spanning the years 2010 to 2021. Economic data are culled from the statistical yearbooks and public reports of the corresponding municipalities for the same temporal range.

## 6. Analytical Outcomes

### 6.1. Temporal Evolution Analysis of Water Resource Systems and High-Quality Economic Indices

Utilizing Equations (1) and (2), the indices for water resource systems and high-quality economic development within the Yangtze River Delta from 2010 to 2021 were computed. The results for the years 2010, 2015, and 2021 are delineated in Figures 1 and 2. As evinced by Figure 1, the mean value of the water resource system indices across the various jurisdictions exhibited a sustained upward trajectory, ascending from 0.24 to 0.38—an increment of 0.14, representing a growth rate of 58%. Except for Zhoushan, the water resource system indices for all other cities demonstrated an ascending trend from 2010 to 2015, with the cities of Ningbo, Yancheng, and Shaoxing registering the most substantial increments. In the period between 2015 and 2021, Yangzhou and Nantong saw a marginal decline in their water resource system indices, whereas all other cities maintained their upward momentum, with Zhoushan, Jiaxing, and Ningbo experiencing the most considerable uplift. This principally emanates from the heightened focus of local governments on water resource management, as the implementation of projects aimed at water pollution remediation and environmental construction has augmented the carrying capacity of water resources across these jurisdictions. Concurrently, demographic and economic advancements in certain regions have stimulated increased water consumption requirements.

Figure 2 reveals that the mean indices for high-quality economic development across various regions escalated from 0.31 to 0.53 between 2010 and 2021, marking an increment of 0.22 and a growth rate of 71%. In 2010, only Shanghai boasted a high-quality economic development index surpassing 0.4, with Wenzhou registering the nadir at 0.18. By 2015, this list had expanded to include Nanjing, Hangzhou, Suzhou, Wuxi, Zhoushan, Changzhou, Ningbo, Zhenjiang, and Lishui, all exhibiting an overarching upward trajectory, with Wenzhou, Suqian, Suzhou, and Nanjing manifesting particularly pronounced growth. By 2021, every city within the Yangtze River Delta had a high-quality economic development index exceeding 0.4, signifying a notable elevation in the quality of economic development.

Historically, the region has been a magnet for substantial foreign capital and an influx of transient populations, thereby catalyzing urban population growth and the expansion of construction land. Concurrently, the provision of public services at the municipal governance level has been commensurately amplified, contributing favorably to high-quality economic growth.

**Figure 1.** Temporal evolution of water resource system indices in the Yangtze River Delta.

**Figure 2.** Time evolution of economic quality development index in the Yangtze River Delta.

### *6.2. Analysis of Spatial Trends in Water Resource Systems and High-Quality Economic Indices*

In order to rigorously delineate the spatiotemporal evolution of water resource systems in conjunction with high-quality economic development within the Yangtze River Delta, we employed Geographical Information Systems (GIS) to plot the standard deviational ellipses and centroidal shifts for these systems for the years 2010, 2015, and 2021. The overarching objective was to scrutinize the spatial equanimity between the two systems (Figures 3 and 4). Figure 3 and Table 3 reveal that between 2010 and 2021, the centroid of water resource quality was predominantly situated in the northeastern region of Huzhou City. The orientation of the ellipse displayed a fluctuating trend, indicating a substantial degree of data dispersion, yet overall, it manifested an upward trajectory. Observationally, the spatial configuration of the water resources the Yangtze River Delta predominantly spans from the northwest to the southeast and substantially encompasses the majority of the central and eastern regions. From the detailed map of the ellipse position, the standard deviation ellipse position moves first to the northwest, then to the southeast, and overall becomes more and more eastward. Although the center of gravity of the ellipse moves to the northwest, the coverage is gradually expanding, indicating that the water resource elements are increasingly concentrated in the east-central region of Jiangsu, Zhejiang, and

Shanghai, especially in the eastern coast of Wenzhou, Taizhou, Zhoushan, and so on, in the environmental quality of the outstanding performance makes the ellipse more skewed to the southeast and the coverage of a larger area.

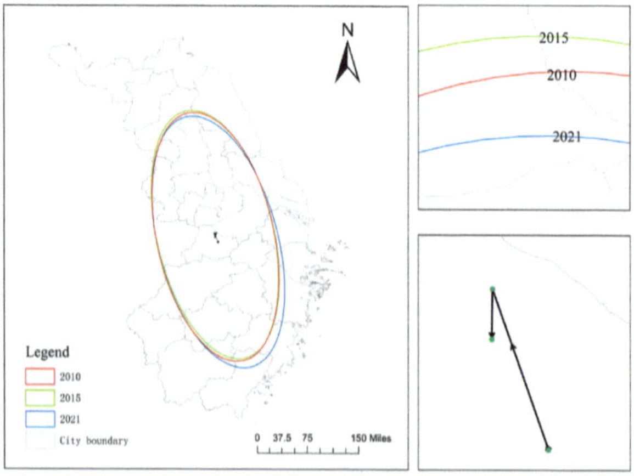


**Figure 3.** Spatial distribution of water resource systems in the Yangtze River Delta.

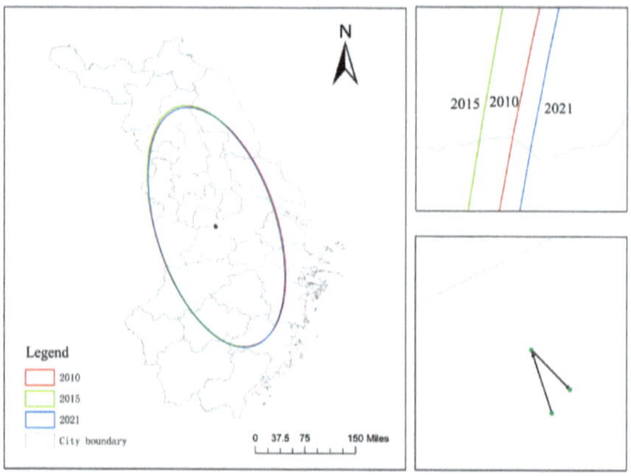


**Figure 4.** Spatial distribution of high-quality economic development in the Yangtze River Delta.

**Table 3.** Parameters of standard deviation ellipses for water resources systems and high-quality economic development in the Yangtze River Delta.

| Index | Year | Centroid Coordinates | | Standard Deviation along X-Axis | Standard Deviation along Y-Axis | Azimuth Angle |
|---|---|---|---|---|---|---|
| | | Longitude | Latitude | | | |
| Water Resources System | 2010 | 119°59′2.612″ | 30°55′49.256″ | 2.650929 | 1.273938 | 166.76 |
| | 2015 | 119°59′5.183″ | 30°58′50.29″ | 2.655153 | 1.271567 | 165.64 |
| | 2021 | 120°24′9.608″ | 30°49′10.88″ | 2.703458 | 1.301092 | 164.38 |
| High-Quality Economic Development | 2010 | 120°1′36.44″ | 31°9′45.907″ | 2.61356 | 1.33335 | 160.94 |
| | 2015 | 120°0′38.563″ | 31°10′43.28″ | 2.64105 | 1.308516 | 160.83 |
| | 2021 | 120°1′9.404″ | 31°9′11.851″ | 2.617677 | 1.313448 | 161.20 |

Figure 4 and Table 3 elucidate that between 2010 and 2021, the epicenter of high-quality economic development was principally located in the vicinity of Lake Tai, at the confluence of Huzhou, Suzhou, and Wuxi, without any substantial alterations. The orientation of the ellipse remained relatively stable. Like the water resource system, the economic development ellipse also predominantly spans from the northwest to the southeast and substantially covers the central and eastern regions of the Yangtze River Delta. This suggests that the high-quality economic development has not led to egregiously disparate regional disparities. According to the detailed map, the ellipse of high-quality economic development is slowly moving eastward, and the center of gravity is shifting eastward as well. Further analysis ascertains that the southeastward inclination is correlated with the gravitational pull of technological talent in Shanghai and Suzhou, and is closely related to the integrated development demonstration areas formed in Suzhou Wujiang, Jiaxing Jiashan, and Shanghai Qingpu. Conversely, the northwestward inclination is associated with the elevated levels of economic and societal development in Nanjing, the provincial capital.

### 6.3. Coupling and Coordination Degree of Water Resource Systems and High-Quality Economic Development: Dynamic Changes over Time

According to the calculated results of the coordination degree of water resource systems and high-quality economic development in the Yangtze River Delta from 2010 to 2021, an analysis is conducted in conjunction with the classification criteria in Table 2 to assess the harmonious development between water resource systems and high-quality economic development (Figure 5). Between 2010 and 2015, the indices of coordination degree across various regions ranged from 0.2 to 0.7, encompassing categories such as moderate discoordination and decline, mild discoordination and decline, moderate harmonious development, and optimal harmonious development, indicating an overall trend toward beneficial progress. In 2010, areas exhibiting superior development included cities such as Shanghai, Hangzhou, Lishui, and Nanjing. This was partly attributed to the rapid economic growth in these cities, which placed a greater emphasis on economic harmonious development. Additionally, cities such as Lishui and Jinhua, with abundant water resources and high developmental carrying capacities, were conducive to coordinated development between the economy and water resources. By 2021, Shanghai and Hangzhou emerged as cities exemplifying exceptional coordinated development, with coordination degrees exceeding 0.7, while other cities all maintained coordination degrees greater than 0.5, falling within the realm of either moderate or optimal harmonious development.

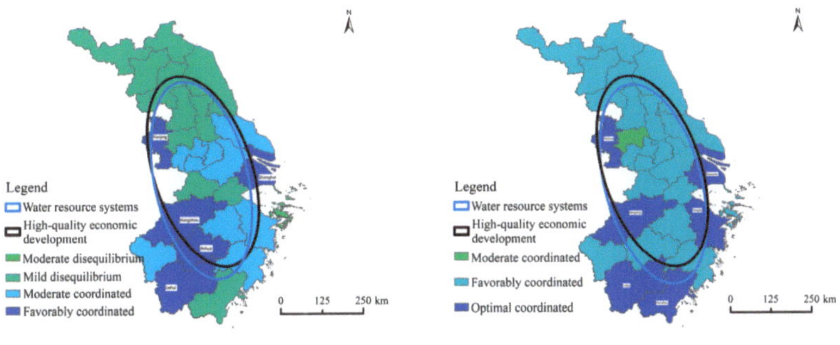


**Figure 5.** Degree of development and spatial ellipse of the coupling between water resource systems and high-quality economic development in the Yangtze River Delta ((**left**) 2010, (**right**) 2021).

According to Figure 6, which illustrates the relative development indices, the mean relative development degree for high-quality economic growth and water resource systems across various regions increased from 1.32 to 1.50 between 2010 and 2021—an increment of 0.18, representing a 14% growth rate, indicative of a pioneering trajectory in regional high-quality development. Within the Yangtze River Delta, advancements in economic prosperity, population growth, and societal evolution have precipitated an escalation in both domestic and industrial water consumption, thereby exerting considerable strain on the water resource systems and complicating their harmonious integration with economic objectives. Presently, while sustaining high-quality economic development, it is imperative to prioritize the conservation and judicious utilization of water resource systems to avert another severe imbalance with water resource carrying capacities and to foster a synergetic and efficient equilibrium between high-quality economic development and water resource utilization.

**Figure 6.** Relative development of water resource systems and high-quality economic development in the Yangtze River Delta.

## 7. Conclusions and Discussion

### 7.1. Conclusions

Upon establishing a comprehensive evaluation index system, this study employs the CRITIC method, coupling-coordination models, and standard deviation ellipse models to scrutinize the dynamic coupling and coordination relationship between water resources and high-quality economic development across 25 cities in the Yangtze River Delta. The research ascertains that both the integrated utilization of water resources and the level of high-quality economic development in the region are on an upward trajectory. The coupling and coordination status has undergone a transition from moderate discoordination to optimal coordination, with the coupling type evolving from a medium-level adjustment to a high-level coordinated state. The epicenter of high-quality economic development is concentrated around Lake Taihu, particularly at the confluence of Huzhou, Suzhou, and Wuxi cities, exhibiting negligible shifts and maintaining a stable azimuth angle oriented from northwest to southeast, essentially enveloping the central and eastern portions of the Yangtze River Delta. The centroid of water resource system quality is located in the northeastern part of Huzhou City. Its azimuth angle manifests fluctuating tendencies, and the data exhibit a significant dispersion. Overall, the trend is ascending, and the spatial configuration is also oriented from northwest to southeast, broadly covering most areas in the central and eastern parts of the Yangtze River Delta.

### 7.2. Discussion

The issue of the relationship between the regional economy and the environment is an important part of the strategic topics such as the new normal of the Chinese economy, the construction of new urbanization, the construction of ecological civilization, and the new height of environmental protection in the new era. Currently, there are fewer studies on spatio-temporal patterns at small scales (prefectural municipalities and the following administrative units) based on large scales (large watersheds and national strategic regions); therefore, this paper selects data from 25 prefectural municipalities in Yangtze River Delta to measure the spatio-temporal patterns at small scales. The limitation of this paper is that it does not explore the relationship between the economic and environmental systems and their internal components of the subsystems, and only reflects the degree of coordination between the two systems from the numerical law, and also does not explore the many factors affecting the environmental system. In addition, in the selection of indicators for constructing the system, we consider the availability of data and based on previous research to screen out all of the indicators, in which there are inevitably oversights. In this paper, we have only conducted some preliminary explorations on economic and environmental issues in Jiangsu, Zhejiang and Shanghai, and we should strengthen further research on urbanization, economic development, ecological, and environmental effects of high-quality development, and refined measurement of environmental pollution and environmental quality in the future.

To harmonize the development of regional water resources with economic and social advancements, the following measures could be intensified. First, to carry out special remedial actions for the management of water abstraction, comprehensively carry out the verification and registration of water abstraction outlets, grasp the number of constructed water abstraction outlets, compliance and monitoring and measurement of the status quo, standardize the behavior of water abstraction, and improve the regulatory mechanism of water abstraction outlets, so as to lay a solid foundation for the management and control of water use. In areas where the verification and registration tasks have been completed, it is necessary to categorize and rectify the problems, and promote the development and utilization of water resources in the basin to improve the order significantly. Second, to solve the water problems in the Yangtze River Delta, we must grasp the crux of the problem, and implement the requirement of "making water resources a rigid constraint" without compromise, and adjust human behavior and correct human misbehavior in a timely manner through strong regulation, so as to promote upgrading and upgrading of industries, and advance the high-quality development of the economy and society. Third, strict water resource demonstration and water permit management, water resource conditions as an important constraint on the approval of relevant planning and construction projects, effectively strengthen the demonstration of water resource and water permit management, regulate the use of water in accordance with the law, and strictly supervise the aftermath of the incident. To promote the zonal management of water resource development and utilization, based on the evaluation of the carrying capacity of water resources, study the establishment of a zonal management system for the development and utilization of water resources, accurately identify basins and regions where water resources are over-exploited, moderately exploited, and under-exploited, and implement differentiated regulatory policies, so as to improve the level of management refinement, and to better facilitate the balancing of the population, urban and industrial development with water resource conditions. Fourthly, give play to the incentive and spurring role of the most stringent water resource management system assessment. Further improve the assessment content, optimize the assessment indicators, improve the assessment mechanism, and give greater play to the role of the assessment baton. Further strengthen the daily inspection, take random checks, unannounced visits, and other ways, strict implementation of supervision and inspection, to promote the local and relevant units to fulfill their duties in accordance with the law, and improve the management capacity and level. We should take warning and disciplinary actions as the guide, and urge immediate rectification of the problems found, and resolutely investigate, punish, and hold accountable those with serious problems.

**Author Contributions:** Conceptualization and methodology, G.Y.; resources and writing, D.L.; project administration and funding acquisition, Q.D. All authors have read and agreed to the published version of the manuscript.

**Funding:** This research received no external funding.

**Data Availability Statement:** The data used in this paper are all from the statistical data officially released by China.

**Conflicts of Interest:** The authors declare no conflict of interest.

## References


1. Hashimoto, T.; Stedinger, J.R.; Loucks, D.P. Reliability, resiliency, and vulnerability criteria for water resource system performance evaluation. *Water Resour. Res.* **1982**, *18*, 14–20.
2. Dee, N.; Baker, J.; Drobny, N.; Duke, K.; Whitman, I.; Fahringer, D. An environmental evaluation system for water resource planning. *Water Resour. Res.* **1973**, *9*, 523–535.
3. Hitchcock, R.K. Water resource use and management in the Okavango system of southern Africa: The political economy of state, community and private resource control. *Botsw. Notes Rec.* **1999**, *31*, 83–92.
4. Kalhor, K.; Emaminejad, N. Sustainable development in cities: Studying the relationship between groundwater level and urbanization using remote sensing data. *Groundw. Sustain. Dev.* **2019**, *9*, 100243.
5. Hall, B.; Currell, M.; Webb, J. Using multiple lines of evidence to map groundwater recharge in a rapidly urbanizing catchment: Implications for future land and water management. *J. Hydrol.* **2020**, *580*, 124265.
6. Minnig, M.; Moeck, C.; Radny, D.; Schirmer, M. Impact of urbanization on groundwater recharge rates in Dübendorf, Switzerland. *J. Hydrol.* **2018**, *563*, 1135–1146.
7. Pienaar, A.; Brent, A.C.; Musango, J.K.; De Kock, I.H. Water resource infrastructure implications of a green economy transition in the Western Cape Province of South Africa: A system dynamics approach. *S. Afr. J. Ind. Eng.* **2017**, *28*, 78–94.
8. Salerno, F.; Gaetano, V.; Gianni, T. Urbanization and climate change impacts on surface water quality: Enhancing the resilience by reducing impervious surfaces. *Water Res.* **2018**, *144*, 491–502.
9. Ren, B.; Sun, B.; Shi, X.; Zhao, S.; Wang, X. Analysis of the cooperative carrying capacity of Ulan Suhai Lake based on the coupled water resources–water environment–water ecology system. *Water* **2022**, *14*, 3102.
10. Luo, Z.; Zuo, Q. Evaluating the coordinated development of social economy, water, and ecology in a heavily disturbed basin based on the distributed hydrology model and the harmony theory. *J. Hydrol.* **2019**, *574*, 226–241.
11. Jia, S.; Zhang, S.; Yang, H.; Jun, X. Relationship between industrial water use and economic development—Kuznets curve of water use. *J. Nat. Resour.* **2004**, *19*, 279–284.
12. Fu, M. Environmental Kunets Curve in China: Shape, Inflection Points and Influencing Factors. *Quant. Tech. Econ. Res.* **2008**, *25*, 40–55.
13. Tong, H.; Yang, Y.; Wang, J.; Feng, Y. Outlook of China's green economy development—Scenario analysis based on system dynamics model. *China Soft Sci.* **2015**, *6*, 20–34.
14. Rueff, H.; Rahim, I.U.; Kohler, T.; Mahat, T.J.; Ariza, C. Can the green economy enhance sustainable mountain development? The potential role of awareness building. *Environ. Sci. Policy* **2015**, *49*, 85–94.
15. Song, D.; Ma, Y. Research on the coupling of green water utilization rate and economic high-quality development in the Yellow River Basin—Taking Henan Province as an example. *Ecol. Econ.* **2021**, *37*, 14–19+42.
16. Xiong, Y.; Su, M.; Zhang, F.; Li, J. Study on the relationship between urbanization process and water resources use response in ChangZhuTan city cluster. *Hum. Geogr.* **2018**, *33*, 69–76. [CrossRef]
17. Han, R.; Tong, L.; Tong, W.; Yu, J. Progress and review of research on the relationship between economic and environmental development. *China Popul. Resour. Environ.* **2012**, *22*, 119–124.
18. Shi, Y.; Zhu, X.; Sun, X.; Li, Y.; Wei, T. Dynamic evaluation of coordinated development of regional economic-environmental composite ecosystems: Taking Lianyungang as an example. *J. Ecol.* **2010**, *30*, 4119–4128.
19. Di, D.; Wu, Z.; Guo, X.; Lv, C.; Wang, H. Value stream analysis and emergy evaluation of the water resource eco-economic system in the Yellow River Basin. *Water* **2019**, *11*, 710. [CrossRef]
20. Lv, S.; Ma, Y.; Ye, J.; Li, Q. Study on the quantitative relationship between urbanization and water resources utilization in the Central Plains city cluster. *J. Irrig. Drain.* **2016**, *35*, 7–12. [CrossRef]
21. Ren, B.; Du, Y. Coupled synergistic relationship of economic growth-industrial development-ecological environment in the Yellow River Basin. *China Popul. Resour. Environ.* **2021**, *31*, 119–129.
22. Fan, J.; Wang, Y.; Wang, Y. Research on regional high-quality development based on geographic units—Another discussion on the difference in conditions and focus of development between the Yellow River Basin and the Yangtze River Basin. *Econ. Geogr.* **2020**, *40*, 1–11.
23. Zhang, L.; Qiu, S.; Yan, L.; Du, J.; Shen, H. Analysis of coupled coordination between water resources utilization and economic and social development in Henan Province based on basin perspective. *Reg. Res. Dev.* **2022**, *41*, 14–19.
24. Jiao, W.; Min, Q.; Li, W.; Anthony, M.F. Water ecological carrying capacity based on ESEF: Theory, modeling and application. *J. Appl. Ecol.* **2015**, *26*, 1041–1048.
25. Liu, S.; Chen, J. Research on water resources carrying capacity based on neural network theory. *Resour. Sci.* **2007**, *9*, 99–105.
26. Yang, L.; Ren, Y.; Wang, Z.; Hang, Z.; Luo, Y. Simulation and Economic Research of Circulating Cooling Water Waste Heat and Water Resource Recovery System. *Energies* **2021**, *14*, 2496. [CrossRef]
27. Xue, J.; Chen, W.; Zhang, L. Measurement and Spatial Pattern of Comprehensive Urbanization Level in Chinese Municipalities. *Econ. Geogr.* **2010**, *30*, 2005–2011.
28. Deng, H.; Yang, L.; Pan, X. Whether the opening of high-speed railroad can help industrial structure upgrade: Facts and mechanisms. *Financ. Res.* **2020**, *46*, 34–48.
29. Xu, L.; Huang, Y.; Liu, A. Study on water resources carrying capacity of Jiangsu Province based on principal component analysis. *Yangtze River Basin Resour. Environ.* **2011**, *20*, 1468–1474.

30. Duan, X.; Luan, F. Evaluation of water resources carrying capacity in Xinjiang based on fuzzy comprehensive judgment. *China Popul. Resour. Environ.* **2014**, *24* (Suppl. S1), 119–122.
31. Su, W.; Wang, C. Research on the evaluation of urban water resources carrying capacity in Sichuan Province based on DEA. *J. Southwest Univ. Natl. (Humanit. Soc. Sci. Ed.)* **2015**, *36*, 116–119.
32. Cao, Z.; Dong, H.; Han, Y. Research on the mechanism of industrial agglomeration on total factor productivity and regional heterogeneity. *Soft Sci.* **2020**, *34*, 50–58. [CrossRef]
33. Zheng, D.; Xu, W.; Jiang, J.; Lv, L. Analysis of the evolutionary trend and coordinated development of water resources carrying capacity and urbanization quality in China. *Econ. Geogr.* **2021**, *41*, 72–81. [CrossRef]

*water*

MDPI

*Article*

# Sedimentation Characteristics of the Fluctuating Backwater Area at the Tail of Cascade Reservoirs: A Case Study of the Three Gorges Reservoir

**Jun Li [1], Hua Ge [2,*], Yanrong Ping [1], Xianyong Dong [3], Lingling Zhu [4] and Yaochang Ma [1]**

[1] Upper Changjiang River Bureau of Hydrological and Water Resources Survey, Chongqing 400021, China; cqmayc@139.com (Y.M.)
[2] Changjiang River Scientific Research Institute, Wuhan 430010, China
[3] China Three Gorges Construction Engineering Corporation, Chengdu 610095, China
[4] Bureau of Hydrology, Changjiang Water Resources Commission, Wuhan 430010, China
* Correspondence: gehua@mail.crsri.cn

**Abstract:** The construction of cascade reservoirs is associated with considerable uncertainty in sedimentation in the fluctuating backwater area of the terminal reservoir and poses challenges to water safety. The sedimentation characteristics under the influence of multiple factors in the main urban area of the Chongqing river section were analyzed as a case study for the operation of cascade reservoirs in the Jinsha River via the utilization of a large dataset spanning back to the normal storage of the Three Gorges Reservoir. The results of this study indicate that, owing to factors such as upstream water, sediment inflow, reservoir operation, and river sand mining, this river section experienced deposition on the sand bars and erosion in the main channel. The rate of sedimentation increased with sediment inflow, peak flow rate, and duration, while the location of sedimentation shifted as the concentration ratio changed. These results may provide technical support not only for the operation of the Three Gorges Reservoir, but also for the governance of the fluctuating backwater areas of other cascade reservoirs.

**Keywords:** fluctuating backwater area; cascade reservoirs; Three Gorges Reservoir; main urban area of Chongqing

## 1. Introduction

The conditions of water and sediment boundaries in fluctuating backwater areas (FBAs) of reservoirs are often complex as they demonstrate characteristics of both a river channel and a reservoir. This results in notable uncertainty in the evolution of the riverbed [1], which can pose challenges for flood control [2], navigation [3,4], the utilization of sand and stone resources [5], water-related engineering [6], and ecological processes [7–9] in the area. This sedimentation uncertainty has attracted the attention of both scholars and reservoir operation management departments [10,11]. In the upper reaches of the Yangtze River, the Three Gorges Reservoir (TGR) has become the end of a cascade reservoir group, with its water and sediment conditions having undergone significant changes since its design [12,13]. Chongqing is one of the most important cities in China and is located in the FBA of the TGR, meaning sedimentation problems have a greater impact on its economy and society. Therefore, studying its sedimentation characteristics is of vital importance for not only selecting the optimal operation plan, but also fully utilizing the benefits of the reservoir.

Several studies have investigated sedimentation in the TGR, especially in its FBA. Based on observational data before and during the initial storage period of the TGR, analyses indicate that sedimentation primarily occurs in the perennial reservoir area [14,15], while the riverbed morphology is relatively stable between years. Additionally, there is generally no obvious unidirectional erosion and deposition (E&D) in the main urban area of Chongqing [16]. However, some river models indicate that sedimentation will accumulate and negatively impact ports, waterways, municipal infrastructure, and the ecology within this river section [17]. Since the normal storage operation of the TGR in 2008, the sedimentation rate of fine sand has decreased due to the retention of sediment in large reservoirs in the upper reaches of the Yangtze River, resulting in flocculation and sedimentation [18–20]. However, sedimentation is still mainly distributed in the perennial backwater area [21,22], and a lag phenomenon has been observed [23,24]. The tributary estuaries in the reservoir area also have a certain sedimentation amount [25,26]. In the FBA, human activities such as dredging, river regulation, and dock construction have interfered with sedimentation [27]. The usual trend of sedimentation in the flood season and erosion in the dry season has remained unchanged in the main urban area of Chongqing. However, the main sediment transport period has been delayed from the post-flood period to the water level fluctuation period of the following year, with a decreased quantity of sediment transported [28]. A sediment transport analysis has also indicated that there has been a cumulative sedimentation phenomenon in the FBA [29,30]. This may be avoided by increasing the water level dissipation rate of the TGR from the water level in front of the dam at 163 m [31]. In addition, mathematical models have indicated that the sediment saturation coefficient significantly impacts E&D [32], and the influence of the downstream water level leads to random evolution of the riverbed [33]. In the future, under the new water and sediment conditions following the operation of cascade reservoirs in the upper reaches of the Yangtze River, the sedimentation quantity and rate of the TGR will significantly decrease [34], taking up to 560 years to reach equilibrium [35].

The FBAs of cascade reservoirs, particularly tail reservoirs, face more complex water and sediment conditions due to the joint operation of a cascade reservoir [36]. Studies conducted following the design phase of the TGR have demonstrated that the concentrated accumulation of sediment in the FBA would not only lead to the transformation of river types and the elevation of flood season water levels, but also threaten the water depth of port areas and worsen navigation conditions [37]. Currently, studies on the FBA of the TGR, especially the river section in the main urban area of Chongqing, are primarily based on data measured in the initial water storage stage, and measurements of normal water storage in the TGR in 2008 and the operation of the cascade reservoirs in the Jinsha River are lacking. In this study, we elucidated the characteristics of sediment deposition in the FBA of the TGR under the latest water and sediment conditions, utilizing the main urban area of the Chongqing section as a case study. Furthermore, we explored the underlying causes to suggest timely measures to avoid adverse effects. We expect these findings to not only provide technical support for the operation of the TGR, but also for the governance of the FBAs of cascade reservoirs more broadly.

## 2. Materials and Methods

### 2.1. Study Area

Since the normal impoundment of the TGR in 2008, its backwater end has reached the vicinity of Jiangjin (approximately 660 km away from the dam), with an approximately 173.4 km long FBA ranging from Jiangjin to Fuling [38]. The main urban area of Chongqing is located in the FBA of the TGR, with a total length of 60 km. This includes the 40 km section from Dadukou to Tongluoxia in the main stream of the Yangtze River, and the 20 km section from Jingkou to Chaotianmen in the Jialing River. Due to the influence of geological tectonic processes, the river section in the main urban area of Chongqing presents a continuous curved channel shape on the plane. In the FBA of the TGR, there are six continuous bends in the main stream of the Yangtze River and five bends in the Jialing River. The curves are connected by relatively straight transition sections (Figure 1). This river section alternates between wide and narrow, and the shoreline is uneven with protruding stone mouths on the shore. The main stream of the Yangtze River during flood season is generally 700 to 800 m wide, with branching sections measuring up to 1300 m and as little as approximately 300 m (the Tongluoxia section).

During the flood season, the Jialing River section is generally 400 to 500 m wide, reaching up to 800 m and as little as approximately 370 m (the Zengjiayan section).

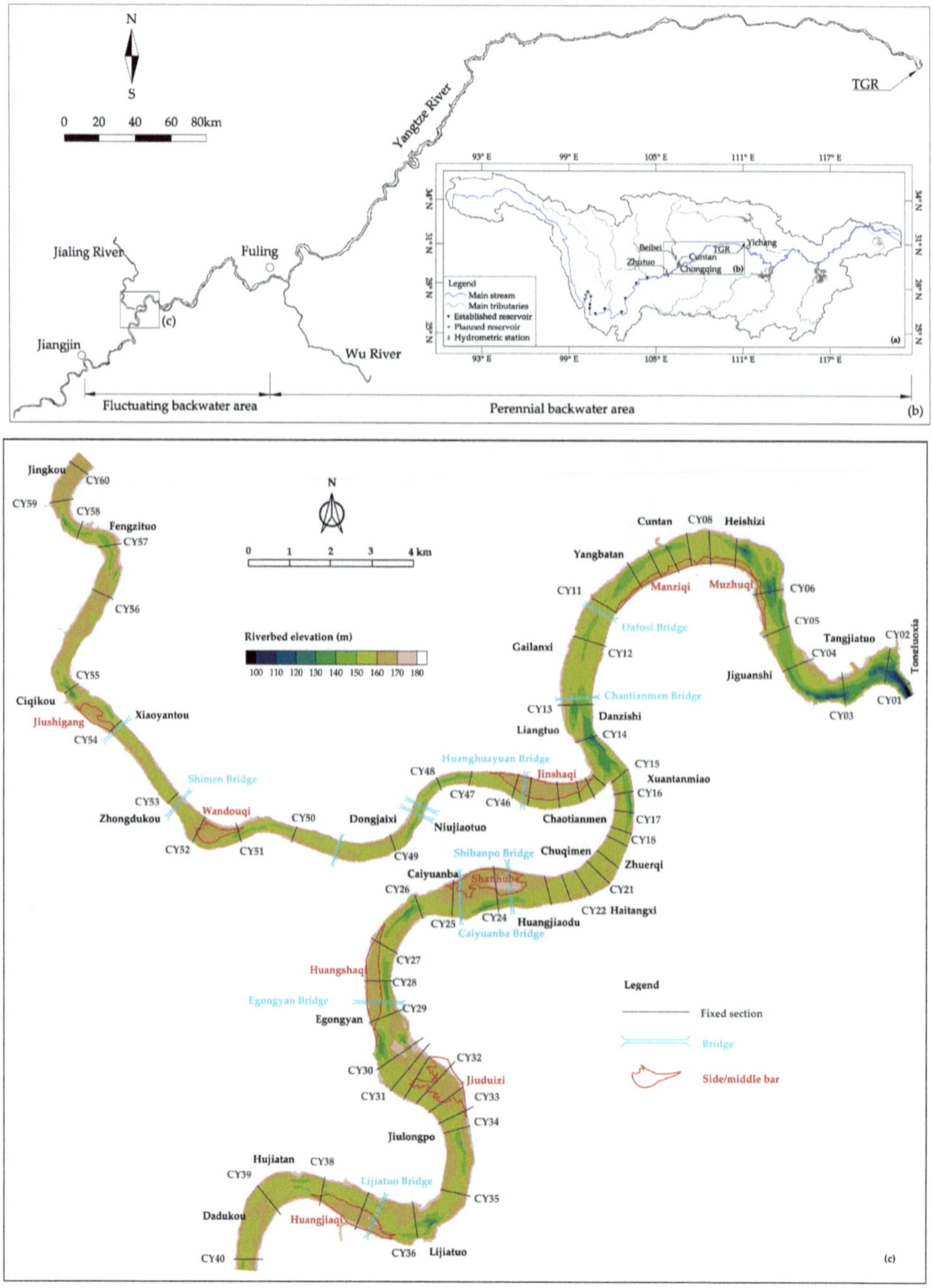


**Figure 1.** Study area: (**a**) the Yangtze River Basin; (**b**) the TGR and its important tributaries; (**c**) the main urban area of the Chongqing river section.

### 2.2. Data

Since 2003, the Hydrological Bureau of the Yangtze River Water Conservancy Commission has conducted annual topographic observations on the main urban area of the Chongqing river section from May to December. A terrain scale of 1:5000 is used in July and December, and 65 fixed sections are observed monthly in the remaining months. From 2010 to 2015, to adapt to changes in the water storage of the TGR, dynamic adjustments were made to the observation tasks. While the terrain observation tasks for the entire river section were discontinued, the number of fixed-section observations and measurements of key river sections were increased. In 2010, a total of 16 observations for fixed sections were made, which were gradually reduced to nine annually by 2015. After 2016, based on the operation of the TGR and the evolving characteristics of the river, further adjustments were made to the observation tasks, with one observation each of the fixed sections before and after flooding, in June and in October, respectively, followed by one topographic observation of the entire river section in December. At the same time, terrain measurement tasks of key river sections were discontinued.

The hydrometric control stations for incoming water and sediment in the main urban area of Chongqing include Cuntan and Zhutuo stations in main stream of the Yangtze River (approximately 152 km upstream of Cuntan) and the Beibei station in the Jialing River (approximately 61 km away from the exit of the Jialing River), all of which have kept extensive records of water and sediment observation data. The data used in this study are detailed in Table 1.

**Table 1.** Data used in this study.

| Type | Period | Measurements | Source |
|---|---|---|---|
| Fixed-section terrain | 2008–2009 | 62, June to November, once a month | Hydrological Bureau of the Yangtze River Water Conservancy Commission |
| | 2010–2015 | 65, decreasing from 16 to 9 times per year | |
| | 2016–2022 | 65, June to October, once a month | |
| Topographic terrain (Entire river section) | 2008–2009 | July, December | |
| | 2016–2022 | December | |
| Topographic terrain (Key river sections) | 2010–2015 | July, December | |
| Flow discharge | 2008–2022 | Daily average | |

### 2.3. Method

Three methods are typically utilized to calculate the quantity of E&D for a river channel. The first method, named the sediment-flux-method, is based on the difference between the sediment flux entering and exiting the river channel. However, this method is only applicable when there are sediment observation facilities at both the river inlet and outlet. For this study area, only the sediment flux exiting the river channel can be obtained at the Cuntan station. The Zhutuo and Beibei stations are too far away from the entrance of the study area to be applicable. Due to the influence of sediment production in the interval, their observed values cannot accurately represent the sediment flux entering the study area. Therefore, this method is not appropriate for this study.

The second method is based on topography. By dissecting the study area, the two measurements above that correspond to sediment storage can be calculated based on topographic data. Although there are several topographical measurements in the study area, their measurement dates are inconsistent with the cross-section. Additionally, the frequency of observations is much smaller than the that of the section terrain. In this study, a third method, based on cross-sectional data [39], is adopted to reflect the entire E&D process. For a given river section, the channel storage (*Vol*) can be expressed by utilizing the frustum volume formula:

$$Vol = \sum \frac{\Delta x\left(A_i + A_{i+1}\sqrt{A_i \cdot A_{i+1}}\right)}{3} \; i = 1, 2, \ldots, ncs - 1 \tag{1}$$

where $A_i$ is the discharge area, $\Delta x$ is the distance from the $i$ to $i$ + 1 section, and $ncs$ is the number of sections. The total difference in the channel storage between two measurements will be the quantity of E&D.

## 3. Results

### 3.1. Volume of Sedimentation

Since the TGR began storing water at a normal water level (175 m) in September 2008, the river section in the main urban area of Chongqing has experienced erosion of 20.672 million $m^3$ of material. Deposition at the side bars totals 2.895 million $m^3$, and scouring in the main channel has reached approximately 23.567 million $m^3$. The process and distribution of E&D are illustrated in Figure 2 and Table 2.

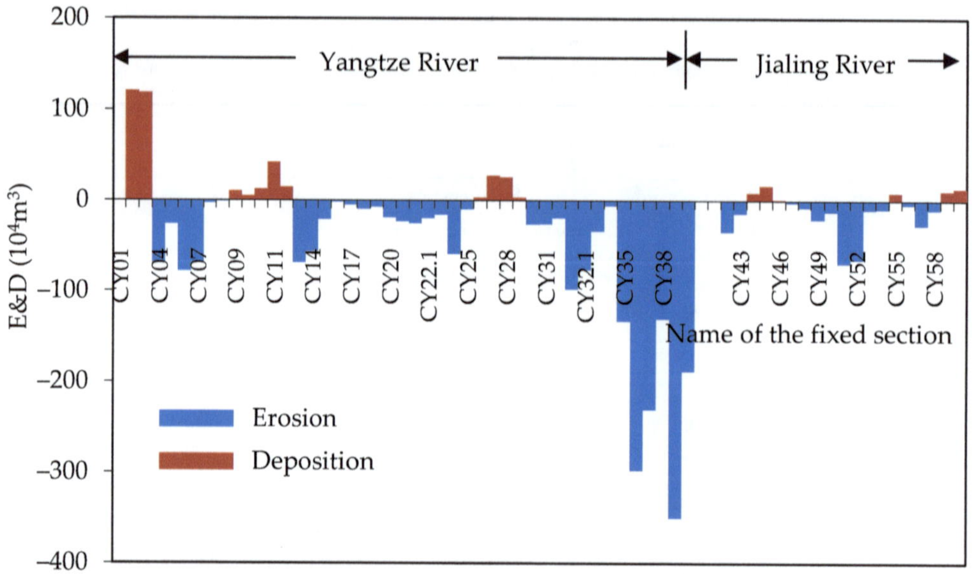


**Figure 2.** Distribution of E&D along the river in the main urban area of Chongqing section since 2008.

By quantifying the distribution of sedimentation, all sections above and below Chaotianmen in the main stream of the Yangtze River, including the Jialing River section, exhibited erosional volumes of 17.568, 0.744, and 2.360 million $m^3$, respectively. The average scouring depths were 1.04 m, 0.07 m, and 0.21 m, respectively (Table 3), and the maximum sedimentation thickness was 18.3 m in the CY02 section. They were located on the left side of the Tangjiatuo area, approximately 14 km below the confluence, with an elevation of approximately 147 m after sedimentation (Figure 3).

### 3.2. Plane Changes of Shorelines and Sand Bars

The main urban area of the Chongqing section is a mountainous river with steeply sloped banks on both sides. No significant changes to the E&D of the 170 m shoreline were observed between 2008 and 2022. Although some areas had been adjusted due to urban construction, the plane changes were relatively small, mostly within 30 m, and the shoreline was relatively stable. The plane changes of the shoreline in the main urban area of Chongqing section are depicted in Figure 4.

**Table 2.** E&D in the main urban area of Chongqing section since the normal storage of the TGR in 2008, unit: $10^4$ m.

| Time Interval | Main Stream of the Yangtze River | | Jialing River | Whole | Notes |
|---|---|---|---|---|---|
| | Below Chaotianmen | Above Chaotianmen | | | |
| September 2008~December 2008 | −37.4 | −24.6 | −66.8 | −128.8 | Storage period in 2008 |
| December 2008~June 2009 | −33.5 | −73.7 | −18.2 | −125.4 | Decline period in 2009 |
| June 2009~September 2009 | −59.9 | 42.6 | 57 | 39.7 | Flood season of 2009 |
| September 2009~November 2009 | 41.6 | −47.1 | −72.2 | −77.7 | Storage period in 2009 |
| November 2009~June 2010 | 16.1 | 70.4 | 94.3 | 180.8 | Decline period in 2010 |
| June 2010~September 2010 | 70.9 | 43 | −154.3 | −40.4 | Flood season in 2010 |
| September 2010~December 2010 | 43.8 | 22 | 139.3 | 205.1 | Storage period in 2010 |
| December 2010~June 2011 | −113.6 | −84.8 | −65.9 | −264.3 | Decline period in 2011 |
| June 2011~September 2011 | −28.9 | 29.7 | 16.8 | 17.6 | Flood season in 2011 |
| September 2011~December 2011 | 12.5 | 53.8 | 19.4 | 85.7 | Storage period in 2011 |
| December 2011~June 2012 | −51.4 | −178.1 | −72.6 | −302.1 | Decline period in 2012 |
| June 2012~September 2012 | 166.7 | 30.8 | 91.8 | 289.3 | Flood season in 2012 |
| September 2012~October 2012 | −21.2 | −105.6 | 18.9 | −107.9 | Storage period in 2012 |
| October 2012~June 2013 | 0.4 | −273 | −57 | −329.6 | Decline period in 2013 |
| June 2013~September 2013 | −57.5 | −28.6 | −53.8 | −139.9 | Flood season in 2013 |
| September 2013~December 2013 | −47.6 | −137.3 | 8.1 | −176.8 | Storage period in 2013 |
| December 2013~June 2014 | −80.4 | −151.2 | −78 | −309.6 | Decline period in 2014 |
| June 2014~September 2014 | 108 | 40.2 | −3.3 | 144.9 | Flood season in 2014 |
| September 2014~December 2014 | −89.2 | −238.3 | −7 | −334.5 | Storage period in 2014 |
| December 2014~June 2015 | −37.3 | −160.2 | −53.7 | −251.2 | Decline period in 2015 |
| June 2015~September 2015 | 120.7 | 71.3 | 84.6 | 276.6 | Flood season in 2015 |
| September 2015~December 2015 | −55.1 | −106.8 | −46.6 | −208.5 | Storage period in 2015 |
| December 2015~June 2016 | 67.5 | −21.1 | −43.8 | 2.6 | Decline period in 2016 |
| June 2016~October 2016 | −100.5 | −31 | −1.4 | −132.9 | Flood season in 2016 |
| October 2016~December 2016 | −42.6 | 54 | 22.6 | 34 | Storage period in 2016 |
| December 2016~June 2017 | 25.6 | −112.8 | −17.2 | −104.4 | Decline period in 2017 |
| June 2017~October 2017 | −8.2 | −82.3 | 28.8 | −61.7 | Flood season in 2017 |
| October 2017~December 2017 | 40.4 | 0.1 | −10.4 | 30.1 | Storage period in 2017 |
| December 2017~June 2018 | −37.7 | −164.6 | −41.8 | −244.1 | Decline period in 2018 |
| June 2018~October 2018 | 14.6 | −69.6 | 26.5 | −28.5 | Flood season in 2018 |
| October 2018~December 2018 | −7.7 | −29.1 | 25.4 | −11.4 | Storage period in 2018 |
| December 2018~May 2019 | 2.3 | −101.1 | −40.6 | −139.4 | Decline period in 2019 |
| May 2019~October 2019 | 38 | −67.6 | −24.7 | −54.3 | Flood season in 2019 |
| October 2019~December 2019 | 44.4 | −51.1 | 6.1 | −0.6 | Storage period in 2019 |
| December 2019~May 2020 | 17.9 | 40.8 | 0.7 | 59.4 | Decline period in 2020 |
| May 2020~October 2020 | 48.7 | 148.9 | 118.3 | 315.9 | Flood season in 2020 |
| October 2020~December 2020 | −76.2 | −81.7 | −25 | −182.9 | Storage period in 2020 |
| December 2020~May 2021 | 57.2 | 38.8 | −41.8 | 54.2 | Decline period in 2021 |
| May 2021~October 2021 | −8.3 | 30.1 | 60.8 | 82.6 | Flood season in 2021 |
| October 2021~December 2021 | −11.3 | 2.2 | 26.4 | 17.3 | Storage period in 2021 |
| December 2021~June 2022 | −89.2 | −88.5 | −69.7 | −247.4 | Decline period in 2022 |
| June 2022~December 2022 | 83 | 34.3 | −16 | 101.3 | Flood and storage periods in 2022 |
| September 2008~December 2022 | −74.4 | −1756.8 | −236 | −2067.2 | |

The 155 m and 160 m contour lines were selected to analyze changes in the sand bars of the main urban area of Chongqing. By analyzing the flat change map of the sand bars (Figure 5), it was found that from 2021 to 2022, the Jinshaqi section underwent marked changes, primarily manifesting as the erosion and retreat of approximately 65 m for the

155 m contour line at the head, and an increase of approximately 50 m in sedimentation at the tail estuary. The overall E&D of the other sand bars was relatively small.

**Table 3.** Thickness of E&D in the main urban area of Chongqing section from September 2008 to December 2022.

| Section | | Average (m) | Maximum | |
|---|---|---|---|---|
| | | | Value (m) | Location |
| Main stream of the Yangtze River | Above Chaotianmen | −1.04 [1] | 2.0 | CY34 |
| | Below Chaotianmen | −0.07 | 18.3 | CY02 |
| Jialing River | | −0.21 | 4.4 | CY52 |

Note: [1] "−" indicates erosion.

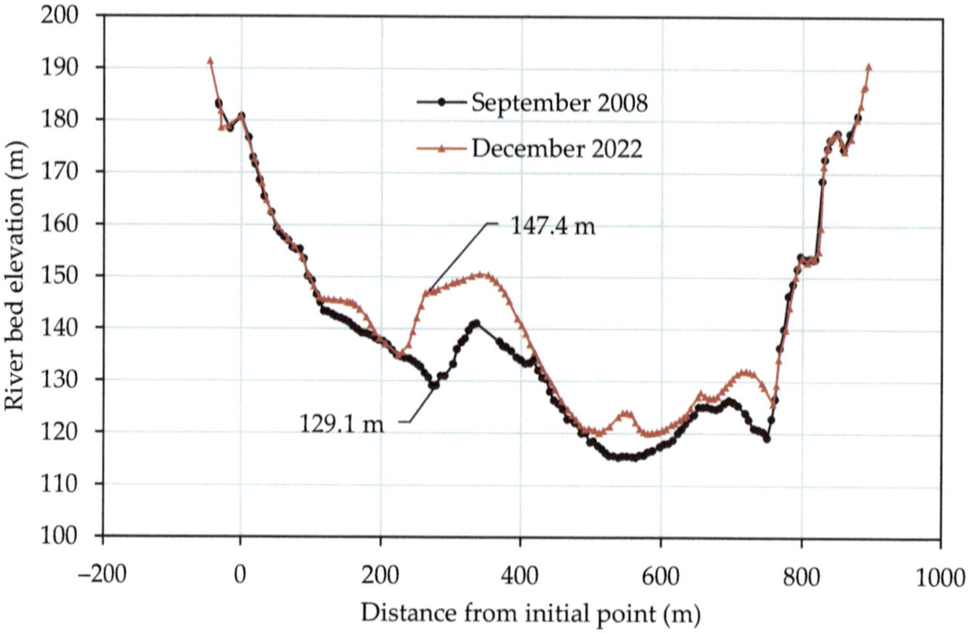


**Figure 3.** Changes in the CY02 section from September 2008 to December 2022.

From 2008 to 2022, the left bank near Tangjiatuo, in the section below the confluence of the main streams of the Yangtze and Jialing rivers, shrank toward the riverbank due to construction, with a maximum shrinkage of approximately 110 m. The section downstream from the Muzhuqi side bar on the right bank near Baishatuo was markedly affected by sand mining, and the front of the bar had retreated notably. The maximum retreating amplitude along the front of the bar was approximately 150 m, and the contour line of 155 m at the top of the bend on the opposite bank had disappeared. The Manziqi side bar on the right bank of the Cuntan section was relatively stable, and its shape and position underwent minimal change. The riverbank on both sides of the section between the Dafosi Bridge and the Chaotianmen Bridge had been adjusted, while the right side bar had been slightly eroded. The contour line of 155 m had retreated by approximately 30 m, and the surface of the left side bar had been slightly eroded. The shoreline in front of the bar was relatively stable. Upstream of the Chaotianmen Bridge, the side bar on the opposite bank of Danzishi was washed away, and the contour line of 155 m had retreated by approximately 150 m. The right bank at the confluence section had been eroded with a maximum retreating amplitude of approximately 70 m for the 155 m contour line.

Above the confluence, the side bar on the left bank of the Zhuerqi section was markedly affected by sand mining, with a retreat of approximately 100 m for the 155 m contour line. There was slight erosion in the middle of Shahuba with a partial retreat of approximately 150 m. The middle section of Xiejiaqi on the right bank, upstream of the Caiyuanba Bridge, had been slightly eroded, retreating by approximately 75 m. The E&D of Huangjiaqi on the left bank was relatively small, and the morphology remained mostly unchanged. Due to factors such as sand mining and construction, the total sand bar area in the Jiulongpo reach was reduced. The surface elevation of the Sanjiaoqi sand bar on the left bank was completely reduced to below 160 m, The elevation of the Jiuduizi sand bar on the right bank underwent similar reductions, as the 160 m contour line at the head had shrunk and retreated by approximately 130 m. The degree of sand mining in the upstream Hujiatan section was relatively high—while the original elevation of the river bottom was mostly above 160 m, the range below 160 m had markedly expanded by the end of 2022.

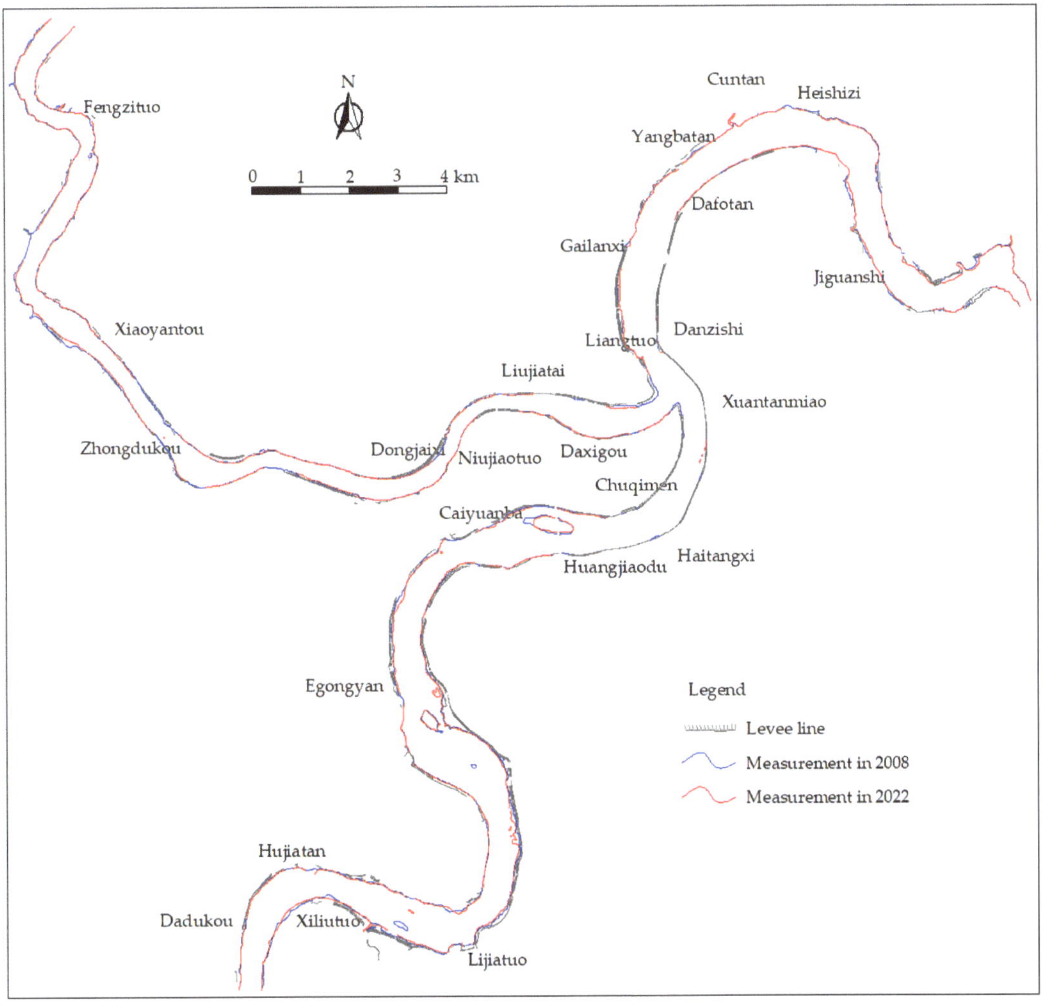


**Figure 4.** Changes in the shoreline (170 m contour line) of the main urban area of the Chongqing river section.

The middle region of the Jinshaqi side bar on the left bank of the mouth of the Jialing River section had eroded, and the 155 m contour line had increased up to 330 m toward the upstream area. The tail had filled with silt, as had the 155 m contour line approximately 60 m toward the river center. The head of the convex bank of the Tuwan section downstream of the Shimen Bridge had retreated by approximately 80 m due to the dredging of the waterway, resulting in sediment accumulating at the top of the bend. The 160 m contour line on the right bank had accumulated silt by approximately 30 m toward the river center, while the shape of the other banks remained relatively stable without notable changes.

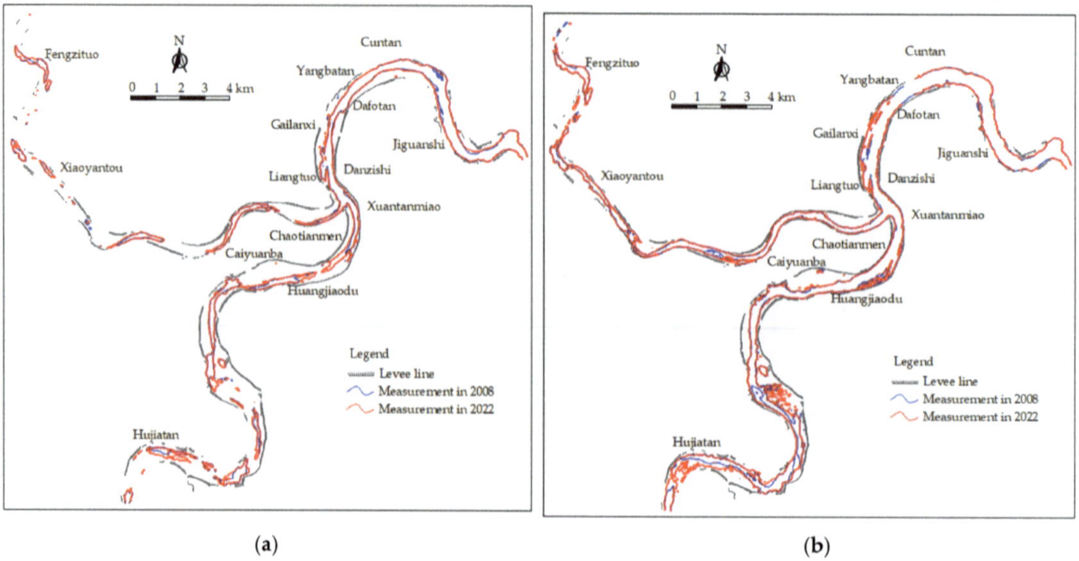


**Figure 5.** Evolution of the sand bar in the main urban area of Chongqing river section: (**a**) the 155 m contour line; (**b**) the 160 m contour line.

### 3.3. Plane Change of Pools

The 140 m and 145 m contour lines were selected to analyze pool changes in the main urban area of the Chongqing section, as depicted in Figure 6. From 2021 to 2022, the overall changes to each pool were relatively small and remained relatively stable.

From 2008 to 2022, deep pools developed in the sections below the confluence, including at the outlet section of Tongluoxia, the opposite bank of Baishatuo, the left side of the Cuntan section, and the Danzishi section. The pools at the exit section developed toward the left bank, with a maximum extension of approximately 180 m for the 140 m contour line and a head drop of approximately 330 m. The Baishatuo section expanded in a transverse direction by approximately 240 m at the head of the deep pool, and by nearly 180 m downstream at the tail of the upper deep pool. The head of the Cuntan deep pool was slightly silted, having retreated by approximately 120 m. The tail of the Danzishi deep pool had been silted to the left and washed to the right, and the left side of the head had been washed, expanding approximately 90 m to the left. Many of these effects were due to sand mining.

Deep pools developed above the confluence on the right side of Xuantanmiao; the opposite bank of Shanhuba, Xiejiaqi, and Huangjiaqi; and the Longfengsi and Jiulongpo areas. The head of the deep pool on the opposite bank of Shanhuba had been slightly silted and had retreated by approximately 40 m, while the head of the deep pool on the opposite bank of Xiejiaqi had been silted and had retreated by approximately 150 m. The length of the deep pool at the 140 m contour line had been shortened from 300 m to 80 m in the upstream area of Jiulongpo, while the other deep pools experienced little change in E&D, with minimal changes to their position and morphology.

The Jialing River section had developed deep troughs on the left side of Zengjiayan, the right side of the upstream area of Niujiaotou, the right side of Zhongshutuo, the right side of Ciqikou, and the right side of Dazhulin. Among them, the sedimentation in the deep trough of Zhongshutuo was more pronounced, with an elevation of approximately 200 m at the end of the deep groove. The width of the tail of the deep groove was reduced from 70 m to approximately 25 m, while the changes in the deep grooves of other sections were relatively small, their positions and morphologies remaining nearly unchanged.

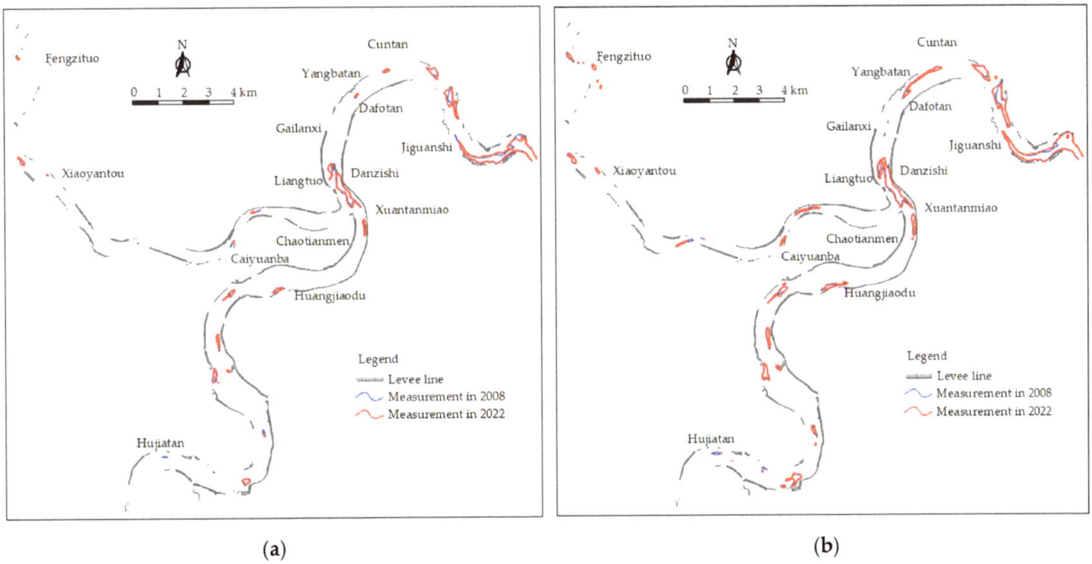


**Figure 6.** Evolution of the pools in the main urban area of the Chongqing river section: (**a**) the 140 m contour line; (**b**) the 145 m contour line.

### 3.4. Changes in the Thalweg

The interannual changes in the longitudinal profile of the thalweg in the main urban area of Chongqing are represented in Figure 7. After the normal impoundment of the TGR in 2008, there was evidence of E&D occurring in the river section, but the overall changes were relatively small, generally within 1 m, except in areas where sand mining and waterway dredging occurred. There are two probable reasons for this: Firstly, most of the deep thalweg in the main urban area of Chongqing is composed of pebbles, which move slowly and exhibit low-intensity sediment transport, resulting in minimal changes to E&D. Secondly, following normal impoundment, this section has generally experienced moderate water and low sand input conditions. Even in 2020, under relatively large water and sand input conditions, deposition and uplift only occurred in local river sections, resulting in no marked impact on the overall characteristics of the thalweg in this river section.

### 3.5. Siltation Distribution

Figure 8 illustrates the distribution of E&D in the main urban area of Chongqing following the normal storage of the TGR below normal water storage levels from July 2008 to December 2022. Both erosion and sedimentation are distributed in the study area. Erosion is primarily distributed in the main stream of the Yangtze River upstream of Chaotianmen and the Jialing River reach upstream of Huanghuayuan Bridge, with an amplitude ranging from 1 m to 3 m. Other areas are mainly characterized by sedimentation, with an amplitude generally within 2 m. In some areas, such as near Tangjiatuo, the amplitude of sedimentation reached greater than 5 m.

Notably, the terrain of some areas exhibits considerable downward cutting, generally exceeding 10 m, occasionally exceeding 20 m. These more extreme terrain changes are due to recent human activities, such as sand mining and waterway dredging.

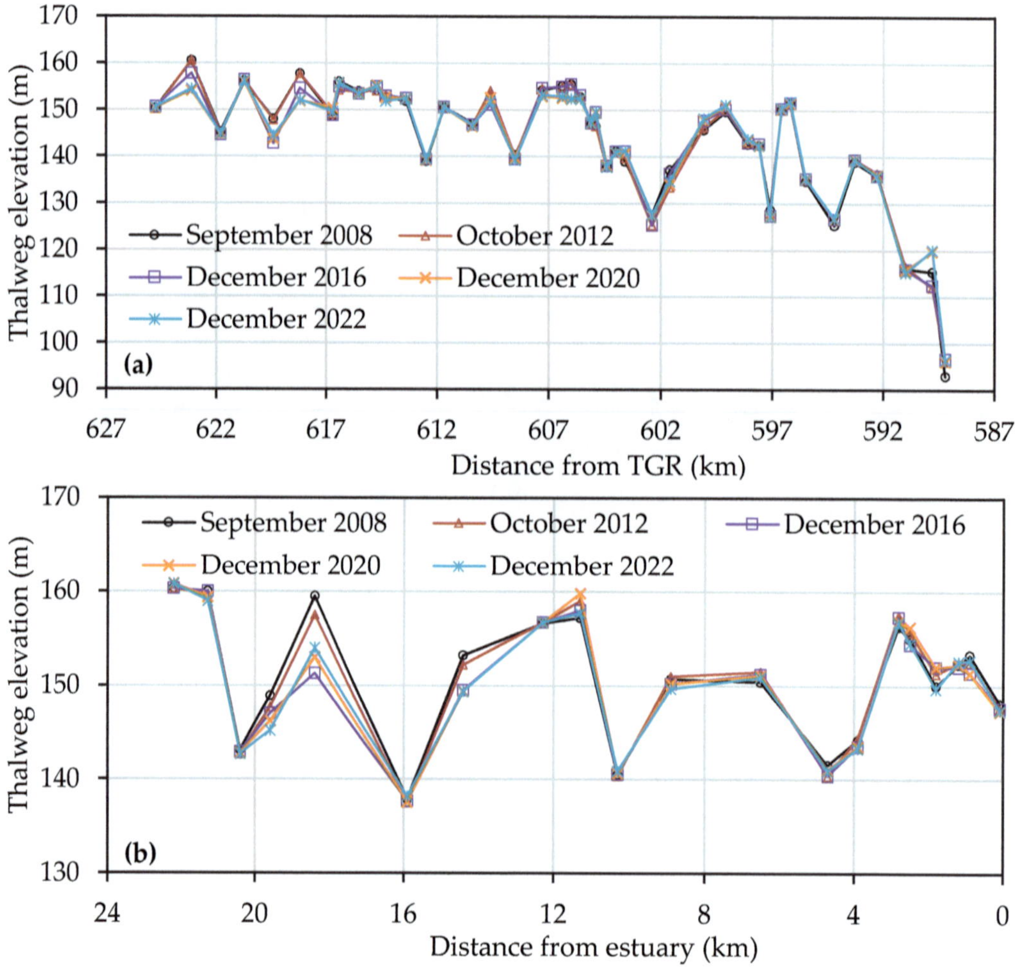


**Figure 7.** Evolution of the thalweg in the main urban area of the Chongqing river section: (**a**) the main stream of the Yangtze River; (**b**) the Jialing River.

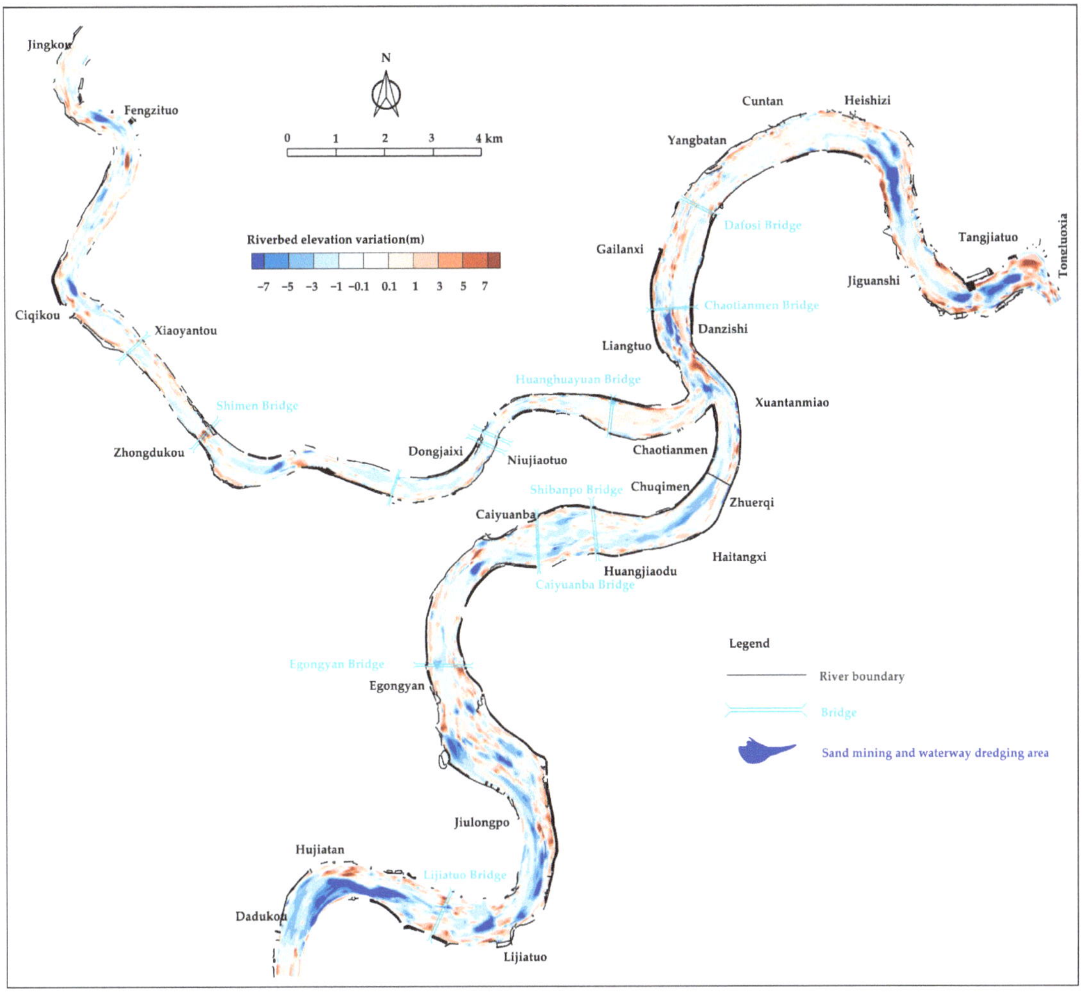


**Figure 8.** Distribution of riverbed elevation changes in the study area from July 2008 to December 2020.

## 4. Discussion

Since the normal impoundment of the TGR, the main urban area of the Chongqing river section has accumulated scouring of 20.672 million $m^3$ due to a variety of factors, including water and sediment inputs, reservoir operational procedures, and river sand mining. This includes depositions of 2.895 million $m^3$ on sand bars and scouring of 23.567 million $m^3$ in the main channel. Overall, between 2008 and 2022, the positions of sand bars and deep pools remained nearly unchanged, with only minor changes in the thalweg of the river. The areas within some river sections that experienced marked changes to E&D were mostly affected by urban construction, waterway dredging, and sand mining. Natural contributions to E&D were relatively small, and the river regime remained fairly stable. We identified three factors, which are discussed below, that have been primarily responsible for the current patterns of evolution in this region.

*4.1. Impact of Income Flow and Sediment*

4.1.1. Total E&D

Figure 9 illustrates the relationship between total runoff, sediment load, and E&D since 2008. The total inflow into the main streams of the Yangtze and Jialing rivers has varied slightly in recent years. However, in 2010, the main streams of the Yangtze and Jialing rivers experienced a total sediment inflow of 223.2 million tons, which was the highest during the study period. Simultaneously, from November 2009 to December 2010, the main urban area of Chongqing experienced deposition of 4.176 million $m^3$ (excluding the impact of sand mining, the same below), which was also the highest during the study period. A relatively large quantity of sand deposition was also observed in 2009, 2012, and 2020, at 2.203, 1.862, and 2.552 million $m^3$, respectively. In 2022, due to relatively low quantities of input sand, 0.492 million $m^3$ of erosion occurred. Similar water and sediment conditions also occurred in the Wujiang River downstream of main urban area of Chongqing, adjacent to the starting point of the fluctuating backwater area [40]. This indicates that when the sediment inflow in the upstream area sharply decreases, sedimentation in the fluctuating backwater area improves [30].

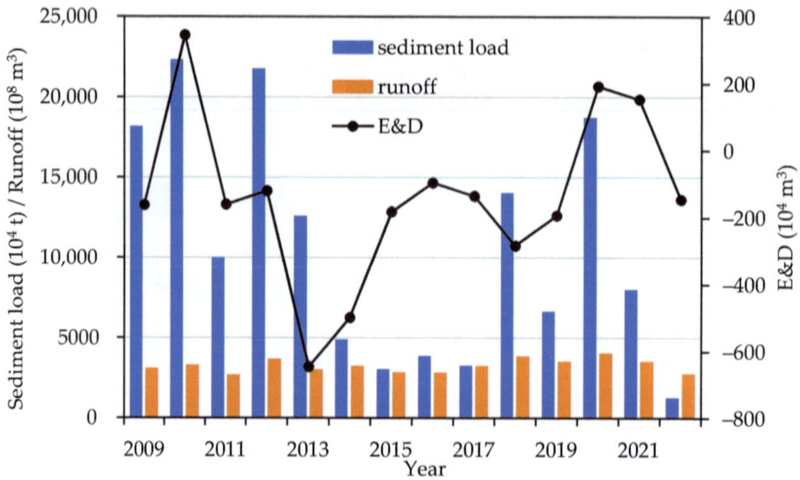


**Figure 9.** Relationship between annual total runoff, sediment load, and E&D.

4.1.2. Process

The impact of incoming water and sediment processes during the year primarily manifests as consistency between the distribution of incoming sediment and flow discharge [41]. The maximum sediment transport capacity in the main urban area of Chongqing occurs when the flow discharge is 12,000–25,000 $m^3/s$ (Cuntan hydrometric station) and is referred to as the main sediment transport flow discharge. If large quantities of sediment are input during this flow discharge, the majority of the sediment is carried away, resulting in a low degree of sedimentation occurring in that river section, accompanied by some scouring. During the flood season, peak flow is typically greater than 25,000 $m^3/s$, the upper limit of the main sediment transport flow discharge. Two distinct patterns are observed in different flood peak conditions:

The larger the flood peak during the flood season, the longer the duration of the flood process, and the more easily sedimentation occurs in the main urban area of the Chongqing section. During the flood season in 2012, from 12 June to 8 August, there were two major floods on 6 July and 24 July, with peak flows of 50,500 and 63,200 $m^3/s$ at the Cuntan hydrometric station. During these two floods, the duration of the flow discharge exceeding 25,000 $m^3/s$ reached 17 days, with 13 and 15 days exceeding 30,000 $m^3/s$, respectively.

Therefore, during the observation period from 12 June to 18 July, and from 18 July to 8 August, a large degree of sediment deposition occurred, with depositions of 2.137 and 2.511 million $m^3$, respectively.

If the peak flow during the flood season is small, the flood process features a sharp peak with a steep rise and fall, a short flood duration, and small quantities of sediment deposited, and erosion is occasionally observed. From 8 August to 8 September 2012, a flood process occurred in the main urban area of Chongqing reach, with a peak flow on September 3rd of 47,300 $m^3/s$ at Cuntan Station. Flows exceeding 25,000 $m^3/s$ only spanned five days. Although the largest sand peak of the year occurred on 6 September, it lagged behind the flood peak by three days, with a corresponding flow rate of 22,900 $m^3/s$. From 8 August to 8 September, not only did the sediment input in the main urban area of the Chongqing section not settle, but a significant transport process was observed, with a total sand loss of 1.755 million $m^3$.

#### 4.1.3. Discharge Ratio of the Main Streams of the Yangtze and Jialing Rivers

The impact of the discharge ratio of the Yangtze and Jialing rivers on E&D in the main urban area of Chongqing primarily manifests as differences in the location of sedimentation [42]. When this ratio is relatively low, the water from the main stream of the Yangtze River imparts a significant lifting effect on the Jialing River section, resulting in the sediment in the Jialing River section being more prone to deposition. For example, from 18 July to 20 August 2014, and from 15 July to 18 August 2015, the discharge ratio of the two rivers was relatively low, at 0.16 and 0.14, respectively. During these periods, the Jialing River section experienced depositions of 0.476 and 0.707 million $m^3$, respectively.

When the ratio is relatively large, the jacking effect of the main stream of the Yangtze River on the Jialing River section is weakened, and the Jialing River section becomes more prone to sediment transport. From 18 July to 17 August 2013, a major flood occurred in the main urban area of Chongqing. This event was primarily induced by rising waters in the Jialing River, which exhibited a maximum annual flow discharge of 24,500 $m^3/s$ on 20 July. On the same day, Cuntan station also experienced a maximum annual flow discharge of 44,900 $m^3/s$. The entire flood process occurred over eight days, with flow discharge durations exceeding 25,000 and 30,000 $m^3/s$ of 12 and 7 days, respectively. During this period, the water inflow from the main stream of the Yangtze River, as measured at the Zhutuo station, was relatively low, with the flow discharge on 20 July measured as only 22,200 $m^3/s$. From 18 July to 25 July, during the flood process, the discharge ratio of the two rivers reached 0.96. The large inflow of the Jialing River resulted in a large degree of erosion in the Jialing River section and the section below Chaotianmen, experiencing sediment erosion quantities of 0.258 and 0.3 million $m^3$, respectively. The river section upstream of Chaotianmen exhibited an accumulated deposition of 1.14 million $m^3$. In 2020, notable flooding events occurred in the main streams of both the Yangtze and Jialing rivers. The average discharge ratio from 22 May to 10 October was 0.28, indicating a strong mutual support effect between the two rivers. During the 2020 flood season, there was marked sedimentation in the main streams of both the Yangtze and Jialing river sections (3.621 million $m^3$).

### *4.2. Impact of the TGR*

Changes in the storage level of the TGR during the storage and dissipation periods after September 2008 are shown in Figure 10. After storage levels reached 175 m, the main urban area of the Chongqing river section was impacted, altering its natural E&D pattern of "flood siltation, dry erosion" to some extent [10]. This manifested primarily as the impact of water storage in the post-flood stage, leading to a decrease in sand transportation, which was unable to carry away pre-flood siltation and the occurrence of sedimentation. Only during the dissipation period, when the storage level decreased and the upstream flow increased, did sand transportation begin, resulting in the river section being eroded.

This river section was weakly affected at the end of the flood season during the early stage of impoundment, especially when the storage level was below 168 m. During this period, the flow measured at Cuntan station was observed to be nearly at the main sediment transport flow discharge status, with the river section still possessing some sediment-carrying capacity. In 2008, 2011, 2012, and 2013, during the initial impoundment period, erosion was observed. Additionally, an equilibrium state of E&D was maintained in 2009. Following the 2010 flood season, there was an autumn flood process, wherein the peak flow measured at Cuntan station reached 33,000 $m^3/s$. Due to the influence of incoming flow, 2.13 million $m^3$ of sediment accumulated in the river section from 10 September to 18 September 2010. As the flood subsided and the main stream returned to its channel, 1.53 million $m^3$ sediment was carried through the reach between 18 September and 30 September. Similarly, from 5 September to 24 September 2014, an autumn flood resulted in slight sedimentation during the initial impoundment period, after which it shifted to erosion.

In the later stage of water storage as the storage level gradually rises, especially when the storage level exceeds 168 m, the river section is primarily influenced by the water level. During this time, flow velocity slows, sediment carrying capacity decreases, and sediment is more prone to deposition. For example, in the later stages of water storage in 2008, 2010, and 2011, notable sedimentation was observed. In the later stage of water storage in 2013, the river section was scoured by 0.86 million $m^3$, which may be primarily due to the influence of sand mining. Calculations based on the cross-section data revealed that terrain changes owing to sand mining amounted to 0.877 million $m^3$. Owing to the impact of sand mining, a slight sedimentation of 0.017 million $m^3$ was observed during this period. In 2014, the river section was also affected by sand mining and exhibited erosion, but also accumulated 0.078000 million $m^3$ of sediment during this period. In 2015, the sediment inflow from the upstream decreased further. Although there was some sedimentation during the flood season, some erosion was observed in the main urban area of Chongqing during the later stage of water storage.

During the pre-flood period, as the storage level declined and upstream flow increased, the river section gradually regained a large amount of sediment carrying capacity. Since 175 m water storage was reached, erosion was the predominant state during the pre-flood period, and in years with more sedimentation during the preliminary stage, the degree of erosion during the pre-flood period was also greater.

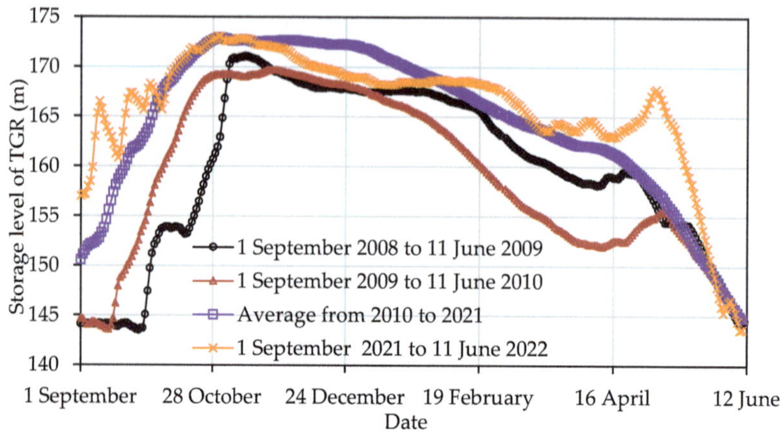


**Figure 10.** Comparison of water level changes in front of the TGR during the storage and dissipation periods from 2008 to 2022.

### *4.3. Impact of Human Activities*

In recent years, sand mining activities have increased in frequency in the main urban area of Chongqing [43], and channel regulation projects have been implemented in multiple places [44,45]. Sand mining and channel regulation have markedly impacted riverbed E&D. Figure 11 highlights typical anthropogenic activities of sand mining and waterway regulation in selected years. From the interannual variation in the cross-section, anthropogenic activities such as these have been observed to result in marked changes to the elevation of local areas of the cross-section, with frequent interannual variations. Since 2011, the Hydrological Bureau of the Yangtze River Commission has organized sand mining surveys on the river sections in the main urban area of Chongqing. According to their survey data, the total quantity of sand mined in 2011 and 2012 was approximately 1.477 and 1.535 million tons, respectively.

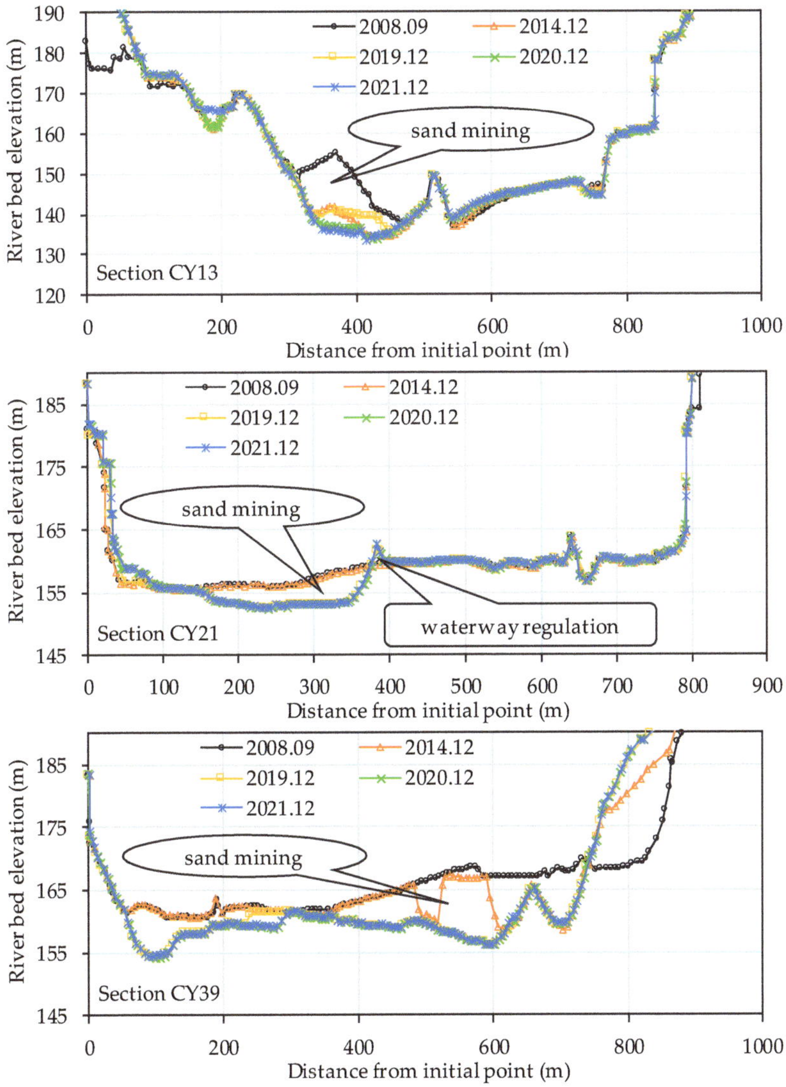


**Figure 11.** Typical cross-sectional changes influenced by human activities.

The impact of sand mining activities on riverbed E&D has two main effects. Firstly, it affects calculations of the quantities of E&D, which masks the natural E&D characteristics of the river. For example, 15 October 2012 to 23 February 2013 experienced the largest volume of sand mined, based on the comparison of fixed cross-section data. By using the channel storage method, we found that the local terrain changes in the main urban area of Chongqing resulting from sand mining amounted to approximately 3.165 million $m^3$. However, the total quantity of E&D in the main urban area of the Chongqing river section was only 3.623 million $m^3$ during this period. Secondly, after sand mining, a large degree of sedimentation was observed in the sand mining area during the flood season. A typical cross-section of sand mining and siltation is shown in Figure 12. The sand mining sections CY02, CY03, and CY06 all experienced large degrees of sedimentation during the flood process. The maximum siltation height in the CY02 section was approximately 5.4 m, and it was 5 m in the CY03 section. The CY06 section also experiences repeated sand mining operations during the year. From December 2011 to June 2012, there were two major sand mining operations near the starting point at 360 and 390 m, with a maximum mining height of nearly 7 m. On 6 July and 24 July, there were two major floods, resulting in severe sedimentation in the excavation area. As of 12 September, the majority of the mining area has been filled with sediment.

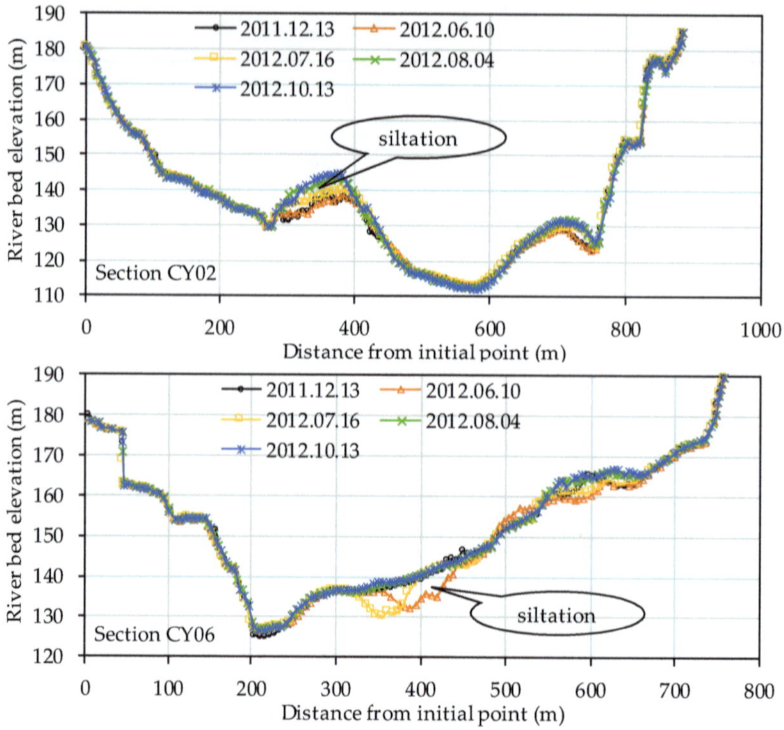


**Figure 12.** Typical cross-section of sand mining and siltation.

## 5. Conclusions

In this study, we analyzed the sedimentation characteristics of the main urban area of the Chongqing river section located in the FBA of the TGR based on terrain measurements and hydrological data collected following the normal storage of the TGR in 2008. The results of this study demonstrate that accumulated erosion has occurred through sedimentation on sand bars and erosion in main channels of 2.895 and 20.672 million $m^3$, respectively.

However, river sections have exhibited little change within 30 m of the shorelines, and the positions of sand bars and deep pools in the river section are nearly unchanged. Furthermore, the amplitude of changes in the thalweg are rarely greater than 1 m.

The evolution of the river sections in this area has been influenced by multiple factors, including water and sediment inputs, reservoir operations, and river sand mining. As sediment inflow and peak flood discharge and duration increase, sedimentation quantities increases, and its location shifts with the change in the confluence ratio of the main streams of the Yangtze and Jialing rivers. In recent years, anthropogenic activities such as sand mining and waterway regulation have led to cumulative erosion.

This study elucidated the sedimentation that has occurred since the normal water storage of the TGR and, by extension, the operation of most cascade reservoirs in the Jinsha River, including those taking place under the water. The results of this study serve as a reference for the evolution of rivers in cascade reservoirs over an extended period into the future. Furthermore, these results provide support for the comprehensive improvement of FBA and the optimization of joint operations in cascade reservoirs.

**Author Contributions:** Conceptualization, J.L. and H.G.; methodology, H.G. and Y.P.; validation, X.D. and L.Z.; formal analysis, J.L. and H.G.; investigation, H.G.; resources, J.L.; data curation, J.L.; writing—original draft preparation, J.L. and H.G.; writing—review and editing, X.D. and L.Z.; visualization, L.Z.; supervision, H.G.; project administration, J.L. and Y.M.; funding acquisition, J.L. and L.Z. All authors have read and agreed to the published version of the manuscript.

**Funding:** This research was funded by the China Three Gorges Corporation (grant number JGAJ0421004 and 0704220) and the Bureau of Hydrology, Changjiang Water Resources Commission (grant number SWJ-CJX23Z10).

**Data Availability Statement:** The data presented in this study are available on request from the corresponding author. The data are not publicly available due to privacy.

**Conflicts of Interest:** Author Xianyong Dong was employed by the company China Three Gorges Construction Engineering Corporation. The remaining authors declare that the research was conducted in the absence of any commercial or financial relationships that could be construed as a potential conflict of interest.

## References

1. Lu, X.; Huang, E.; Wang, S.; Zhao, X.; Guan, X. Bed-load Sediment Transport in Fluctuating Backwater Area of a Steep Slope Reach. *Adv. Eng. Sci.* **2018**, *50*, 89–95. [CrossRef]
2. Zhang, R.; Wu, B.; Liu, S. Cause of Sediment Problems in Flood Seasons of the Niulan River-Dian Lake Water Diversion Project. *J. Basic Sci. Eng.* **2020**, *28*, 1–17. [CrossRef]
3. Deng, Z. Pebble movement law and channel regulation countermeasures in Shangluoqi reach of Three Gorges Reservoir. *Port Waterw. Eng.* **2023**, 120–126. [CrossRef]
4. Su, L. Riverbed evolution and waterway regulation scheme of Guangyangba reach in fluctuating backwater area of the Three Gorges Reservoir. *Port Waterw. Eng.* **2023**, 149–154. [CrossRef]
5. Xia, N.; Xiao, Y.; Hu, X.; Deng, J.; Li, D. Sand mining distribution in fluctuating backwater area of Three Gorges Reservoir. *Port Waterw. Eng.* **2023**, 151–156. [CrossRef]
6. Liu, M.; Zeng, L.; Wu, L.; Zhu, C.; Abi, E. Effect of periodic water-sediment laden flow on damage for steel piles. *Front. Environ. Sci.* **2022**, *10*, 971786. [CrossRef]
7. Xu, G.; Wang, L.; Yang, S.; Yang, W.; Peng, Y.; Sun, Z. Spatial Distribution and Habitat Characteristics of Fish Swarm in Fluctuating Backwater Zone of Three Gorges Reservoir in Impoundment Period. *J. Yangtze River Sci. Res. Inst.* **2021**, *38*, 46–52. [CrossRef]
8. Yang, S.; Xu, G.; Wang, L.; Yang, W.; Xiao, Y.; Li, W.; Hu, J. Field-derived relationships between fish habitat distribution and flow-sediment conditions in fluctuating backwater zone of the Three Gorges Reservoir. *Ecol. Inform.* **2021**, *62*, 101273. [CrossRef]
9. Xiao, Y.; Deng, J.; Yang, S.; Hu, J.; Wang, L.; Li, W. Study on the spawning habitat suitability of four major Chinese carps in the fluctuating backwater area of the Three Gorges Reservoir. *Ecol. Indic.* **2022**, *143*, 109314. [CrossRef]
10. Jianjun, Z.; Cao, H.; Man, Z. Reducing Deposition in TGP Reservoir by Dredging Coarse Sediment. *Sci. Technol. Rev.* **2010**, *28*, 28–36.
11. Hu, C.; Fang, C. Research on sediment problem solutions for the Three Gorges Project and it operational effects. *Sci. Sin. Technol.* **2017**, *47*, 832–844. [CrossRef]
12. Lu, Y.; Zuo, L.; Ji, R.; Liu, H. Deposition and erosion in the fluctuating backwater reach of the Three Gorges Project after upstream reservoir adjustment. *Int. J. Sediment Res.* **2010**, *25*, 64–80. [CrossRef]

13. Tang, X.; Tong, S.; Huang, G.; Xu, G.; Li, X.; Lei, K.; Yao, S. Characteristics of sedimentation and channel adjustment linked to the Three Gorges Reservoir. *Int. J. Sediment Res.* **2021**, *36*, 177–189. [CrossRef]
14. Hu, J.; Yang, S.; Wang, X. Sedimentation in Yangtze River above Three Gorges Project since 2003. *J. Sediment Res.* **2013**, *1*, 39–44. [CrossRef]
15. Li, W.; Yang, S.; Fu, X.; Xiao, Y. Sedimentation characteristics in the Three Gorges Reservoir during the initial operation stage. *Adv. Water Sci.* **2015**, *26*, 676–685. [CrossRef]
16. Peng, W.; Liu, D.; Liu, T.; Wang, X. Analyses on erosion and deposition characteristics of the river reach in Chongqing City. *J. Sediment Res.* **2005**, *6*, 44–50. [CrossRef]
17. Xujing, Z.; Dewei, M.U.; Shiqiang, Z. Sediment Aggregation and Regulation for Chongqing Reach of the Fluctuating Backwater District in Three Georges Reservoir. *J. Chongqing Jianzhu Univ.* **2006**, *28*, 13–17. [CrossRef]
18. Liu, J.; Yang, S.; Shen, Y. Impact of runoff and sediment from the Upper Yangtze River on deposition in the Three Gorges Reservoir. *J. Sediment Res.* **2019**, *44*, 33–39. [CrossRef]
19. Ren, S.; Liu, L. Sediment deposition and countermeasures in the Three Gorges Reservoir. *J. Sediment Res.* **2019**, *44*, 40–45. [CrossRef]
20. Li, W.; Ma, H.; Yang, S.; Yang, W.; Song, C. Sedimentation and One Dimensional Simulation of Fine Sediment in the Three Gorges Reservoir. *J. Basic Sci. Eng.* **2021**, *29*, 251–260. [CrossRef]
21. Li, W.; Li, N.; Yang, S.; Wang, T. Analysis of the sedimentation in the Three Gorges Reservoir based on the sediment carrying capacity. *Adv. Water Sci.* **2016**, *27*, 726–734. [CrossRef]
22. Li, W.; Yang, S.; Xiao, Y.; Fu, X.; Hu, J.; Wang, T. Rate and Distribution of Sedimentation in the Three Gorges Reservoir, Upper Yangtze River. *J. Hydraul. Eng.* **2018**, *144*, 05018006. [CrossRef]
23. Tang, X.; Tong, S.; Xu, G.; Huang, G.; Wang, T. Delayed response of sedimentation in the flood seasons to the pool level of the Three Gorges Reservoir. *Adv. Water Sci.* **2019**, *30*, 528–536. [CrossRef]
24. Li, X.; Ren, J.; Xu, Q.; Yuan, J.; Zhang, W. Impact of cascade reservoirs on the delayed response behaviour of sedimentation in the Three Gorges Reservoir. *J. Geogr. Sci.* **2023**, *33*, 576–598. [CrossRef]
25. Zhu, L.; Xu, Q.; Zhang, O.; Yang, Y.; Wang, W. Sedimentation at estuary of 66 tributaries in the Three Gorges Reservoir. *Sci. Sin. Technol.* **2019**, *49*, 552–564. [CrossRef]
26. Liu, S.; Li, D.; Liu, D.; Zhang, X.; Wang, Z. Characteristics of sedimentation and sediment trapping efficiency in the Three Gorges Reservoir, China. *Catena* **2022**, *208*, 105715. [CrossRef]
27. Wang, B.-Y.; Yan, D.-C.; Wen, A.-B.; Chen, J.-C. Influencing factors of sediment deposition and their spatial variability in riparian zone of the Three Gorges Reservoir, China. *J. Mt. Sci.* **2016**, *13*, 1387–1396. [CrossRef]
28. Yang, Y.; Chen, L.; Huang, J.; Ping, Y.; Liu, T. Study of sediment silting-scouring in Jialing-Yangtze confluence reach during 175 m experimental impoundment period. *Eng. J. Wuhan Univ.* **2017**, *50*, 31–36. [CrossRef]
29. Lei, Y. Study on overall sediment erosion and sedimentation in the fluctuating backwater area of the Three Gorges Reservoir. *Yangtze River* **2018**, *49*, 1–5. [CrossRef]
30. Liu, S.; Zhang, X.; Huang, J. Analysis on recent variation of transport characteristics of gravel bed-load into Three Gorges Reservoir and its impact. *Water Resour. Hydropower Eng.* **2018**, *49*, 82–89. [CrossRef]
31. Zhu, L.; Ge, H.; Dong, B.; Yang, C. Control index of sediment reduction dispatching in Chongqing reach after the storage level of TGR reaching 175 m. *Acta Geogr. Sin.* **2021**, *76*, 114–126. [CrossRef]
32. Ge, H.; Deng, C. 2D Hydraulic modelling of the main urban area of Chongqing, Part 2: Model validation and application. In Proceedings of the 4th International Workshop on Renewable Energy and Development (IWRED), Electr Network, Sanya, China, 24–26 April 2020; IOP: Bristol, UK, 2020.
33. Luo, M.; Yu, H.; Huang, E.; Ding, R.; Lu, X. Two-Dimensional Numerical Simulation Study on Bed-Load Transport in the Fluctuating Backwater Area: A Case-Study Reservoir in China. *Water* **2018**, *10*, 1425. [CrossRef]
34. Ren, S.; Zhang, B.; Wang, W.-J.; Yuan, Y.; Guo, C. Sedimentation and its response to management strategies of the Three Gorges Reservoir, Yangtze River, China. *Catena* **2021**, *199*, 105096. [CrossRef]
35. Chen, P.; Deng, J.; Tan, G.; Lu, J.; Jin, Z.; Zhou, Y.; Shu, C.; Feng, Z.; Tang, R.; Lve, Y.; et al. Prediction research on sedimentation balance of Three Gorges Reservoir under new conditions of water and sediment. *Sci. Rep.* **2021**, *11*, 19005. [CrossRef] [PubMed]
36. Yang, C.; Deng, J.; Qi, Y.; Wang, Y.; Lin, J. Study on Long-term Impact of Small and Medium-sized Floods Regulation on Sedimentation of the Three Gorges Reservoir. *Water Resour. Power* **2020**, *38*, 34–37.
37. Pan, Q. Review of sediment research in Three Gorges Project in recent 60 years. *Yangtze River* **2017**, *48*, 18–22. [CrossRef]
38. Bureau, C.W.R. Yangtze River. Available online: http://slj.cq.gov.cn/ztzl_250/hzzjsst/tzgg_24017/201707/t20170705_6080714_wap.html (accessed on 5 July 2023).
39. Ge, H.; Zhu, L.; Deng, C.; Jiang, L.; Long, R. Decadal Hydro-Morphodynamic Evolution of a Sandy Meander Bend Downstream of the Three Gorges Dam and Impact on Waterway Navigation. *Water* **2022**, *14*, 19. [CrossRef]
40. Lu, Y.; Wang, Y.; Cao, L.; Zhang, Q.; Zhao, D. Fluvial Process of Chongqing Segment of Wujiang River and Its Impacts on Navigation. *J. Yangtze River Sci. Res. Inst.* **2023**, *40*, 186–191. [CrossRef]
41. Huang, R.; Shu, C.; Tang, G. Preliminary Study on Transportation Characteristics of Sediment Concentration Peak of the Three Gorges Reservoir in Flood Season. *J. Basic Sci. Eng.* **2019**, *27*, 1202–1210. [CrossRef]

42. Zhang, F. Influence of confluence ratio on water flow and sediment characteristics in Chongqing of the Yangtze River. *J. Waterw. Harb.* **2023**, *44*, 352–361. [CrossRef]
43. Xiao, Y.; Zhang, S.; Yang, S.; Li, W. Distribution of sand excavation area and simulation of recovery ability for destroyed gravel shoals in upstream of the Yangtze River. *Port Waterw. Eng.* **2020**, 127–131+154. [CrossRef]
44. Chen, X.; Zhang, X. Regulation of the Chongqing section of the Yangtze River in the fluctuating backwater area of the Three Gorges Reservoir. *China Water Transp.* **2008**, 40–41+43. [CrossRef]
45. Fan, S.; Zhu, D.; He, Y.; Li, J. Stepped navigation regulation structures design of Hujiatan beach in fluctuating backwater area of the Three Gorges reservoir. *Port Waterw. Eng.* **2016**, 134–138+157. [CrossRef]

*water*

MDPI

*Article*

# Optimization of Reservoir Level Scheduling Based on InSAR-LSTM Deformation Prediction Model for Rockfill Dams

**Zhigang Fang [1], Rong He [1,*], Haiyang Yu [1], Zixin He [2] and Yaming Pan [1]**

[1] School of Surveying and Land Information Engineering, Henan Polytechnic University, Jiaozuo 454003, China; 212104020042@home.hpu.edu.cn (Z.F.); yuhaiyang@hpu.edu.cn (H.Y.); 212004020049@home.hpu.edu.cn (Y.P.)

[2] Laboratory of Target Microwave Properties, Deqing Academy of Satellite Applications, Deqing 313200, China; yzmleslie@gmail.com

* Correspondence: hnjzhero@126.com

**Abstract:** The Xiaolangdi reservoir has a storage capacity of more than 10 billion cubic meters, and the dam has significant seasonal deformation. Predicting the deformation of the dam during different periods is important for the safe operation of the dam. In this study, a long short-term memory (LSTM) model based on interferometric synthetic aperture radar (InSAR) deformation data is introduced to predict dam deformation. First, a time series deformation model of the Xiaolangdi Dam for 2017–2023 was established using Sentinel-1A data with small baseline subset InSAR (SBAS-InSAR), and a cumulative deformation accuracy of 95% was compared with the on-site measurement data at the typical point P. The correlation between reservoir level and dam deformation was found to be 0.81. Then, a model of reservoir level and dam deformation predicted by neural LSTM was established. The overall deformation error of the dam was predicted to be within 10 percent. Finally, we used the optimized reservoir level to simulate the deformation at the measured point P of the dam, which was reduced by about 36% compared to the real deformation. The results showed that the combination of InSAR and LSTM could predict dam failure and prevent potential failure risks by adjusting the reservoir levels.

**Keywords:** SBAS-InSAR; LSTM; rockfill dam; deformation prediction; reservoir storage level scheduling

## 1. Introduction

Rockfill dams are large hydraulic structures composed of various rock materials equipped with impermeable core walls made of special materials. Dams play important roles in terms of water storage, flood control, sand drainage, and power generation. They have a great influence on the development of industry and agriculture and the local geological environment [1,2].

During the construction of a rockfill dam, different materials are used in different embankment zones; the same embankment zone may be constructed differently in the field, leading to inevitable differences in deformation in different areas of the dam [3]. After the dam is completed and commissioned, the reservoir begins to store water and adjusts its level according to certain rules. The dam body also undergoes seasonal deformation as the hydrostatic pressure on the dam changes with the rainfall and reservoir levels [4].

Rockfill dams have more complex deformation characteristics than conventional structures or concrete dams. To assess the overall structural health in a timely manner, the periodic deformation monitoring of dams is essential. It can be used to predict how a dam will deform over a period of time in the future to determine if any safety hazards exist and to take appropriate actions to prevent the main body of the dam from crumbling or collapsing [5].

Most traditional dam deformation monitoring methods use geodetic techniques, such as using level and total station measurements for observations [6]. In addition, more sophisticated global satellite positioning techniques can be employed [7]. Sensors such as embedded flexible pipes can also be installed inside dams for monitoring [8]. Although these methods are widely used, they are labor-intensive and expensive. In addition, these methods can only reflect the deformation of monitoring points or lines. However, they do not accurately reflect the overall deformation of a dam. Some scholars have used numerical simulations to reconstruct the deformation characteristics of an entire dam using similar methods, provided that the design and construction parameters of each part of the dam are available, and the overall deformation is calculated via computer simulations [9,10]. Currently, many rockfill dams were built and put into operation worldwide so early that some of the dam parameters are missing, and this method cannot be used.

Interferometric synthetic aperture radar (InSAR) can acquire information on the deformation of the Earth's surface on a millimeter scale over time [11]. It has been widely used to map deformations caused by geological events, such as landslides, earthquakes, and volcanic eruptions [12–15]. It has also achieved good results in monitoring the effects of human activities such as mining, groundwater extraction, road subsidence, and urban development [16–24]. In addition, large regional surface studies, such as polar permafrost and glacier monitoring, have great potential [25–27].

With respect to the monitoring of dam deformation using InSAR, Zhou et al. [28,29] showed that the overall deformation of the Shuibuya Dam obtained using the multiple-temporal InSAR (MT-InSAR) technique is in agreement with the results of in situ geodetic measurements. On this basis, the numerical model of dam deformation using finite element analysis can be complemented by surface deformation monitored via InSAR. Ruiz-Armenteros et al. [30] used MT-InSAR to study a dam in southern Spain and observed significant vertical settlement in the main body of the dam, with deformations mainly distributed in the center of the dam and part of the slope adjacent to the dam. Biondi et al. [31] studied the deformation characteristics of the Mosul Dam using the PS-InSAR technique and showed significant settlement at the center of the dam and opposite deformation trends at the two ends of the dam. A crack was found in the dam that seriously threatened its safe operation. Xiao et al. [32] combined ICEsat-2, Sentinel-1, and Sentinel-2 satellite data and used the small baseline subset InSAR (SBAS-InSAR) technique to investigate the failure of the Sardoba Dam in 2020. The dam failure section showed abnormal settlement, and it was hypothesized that seepage may have occurred inside the dam, resulting in changes in the pressure-bearing capacity and structural stability of the dam, thereby causing the accident. Bayik, Abdikan, and Arıkan [33] monitored the Atatürk Dam in Turkey using multiple images from three satellites, ERS, ENVISAT, and Sentinel-1A, in different orbits, and they found that the dam was still settling 28 years after its construction. However, the deformation characteristics of the dam were different in different periods and showed that the water storage level and the dam deformation were not always correlated.

Long short-term memory (LSTM) is one of the most influential methods in the field of deep learning [34–38], and many researchers have used LSTM models to explore the potential information of data in investigations such as surface observations. Jean et al. [39] used an LSTM model to assess seismic vulnerability across India and prioritize areas in need of protective measures. Li et al. [40] achieved real-time monitoring and prediction of landslide displacement using a deep learning framework based on LSTM to study landslides in the Three Gorges area of China. Chen et al. [41] used the LSTM model based on sliding window data to build a model for predicting future precipitation with precipitation data of up to 40 years in some regions of Turkey as the dependent variable and analyzed the lag period of the precipitation time series by adjusting the model's hyperparameter. Radman, Akhoondzadeh, and Hosseiny [42] used the InSAR technique to obtain surface deposition data near Lake Urmia in Iran and developed a predictive LSTM model combining environmental parameters such as groundwater content to investigate the driving factors affecting surface deposition. The LSTM model proved to be excellent for the time series data. We used longer-period InSAR deformation data and combined them with daily water storage level data to better utilize the data mining capability of LSTM.

The overall workflow diagram of this study is shown in Figure 1, and the main work can be divided into three parts:

(1) Obtain the spatial and temporal evolution characteristics of dam surface deformation by constructing a time series model of the Xiaolangdi Dam using SBAS-InSAR.
(2) Analyze the deformation law of the InSAR model and propose an LSTM network model using water storage level data to predict the surface deformation of the dam.
(3) Optimize the prediction model and propose a reservoir level scheduling scheme and finally verify the feasibility of the scheme using the InSAR-LSTM deformation prediction model.

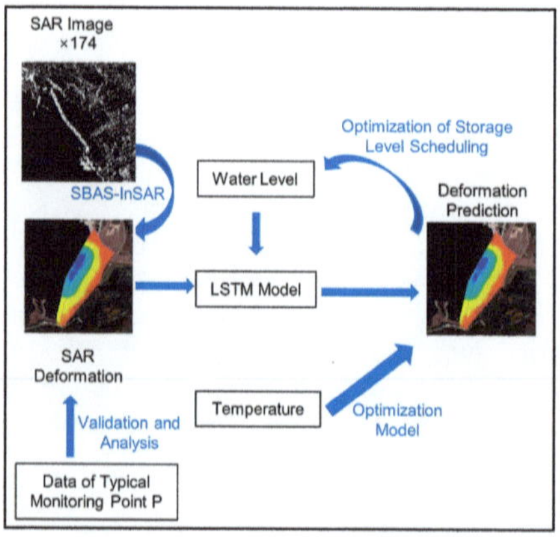


**Figure 1.** The overall workflow diagram of this study.

## 2. Materials

### 2.1. Study Area

The Xiaolangdi Dam was constructed in 2001 and is located 40 km north of Luoyang City, Henan Province, China, on the main stream of the Yellow River, as shown in Figure 2. The dam is a comprehensive large-scale water conservancy project that integrates siltation reduction, flood control, water supply control, irrigation, and power generation. The dam is a key project in the management and development of the Yellow River, controlling a total basin area of 694,000 km$^2$, accounting for 92.3% of the Yellow River.

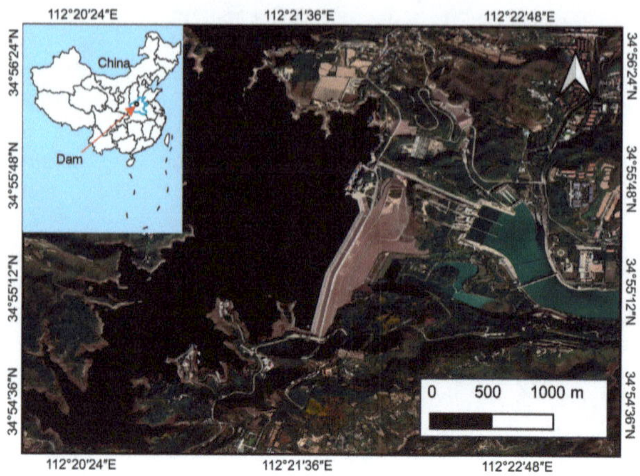


**Figure 2.** Diagram of the location of Xiaolangdi Dam.

Figure 3 shows the structure of the dam. The inside of the dam is a clay-inclined core wall that prevents water seepage, whereas the outside is composed of rock and clay. With a storage capacity of 12.65 billion $m^3$, the highest elevation of the dam is 283 m. The normal storage level is 275 m, the heads above and below the dam are approximately 100 m, and the highest historical water storage level was 273.35 m.

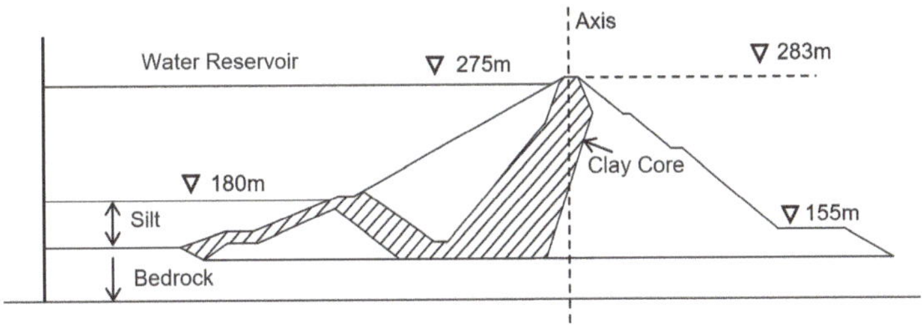


**Figure 3.** Schematic diagram of the main structure of Xiaolangdi Dam.

### 2.2. Dataset

In this study, the radar image data used were from the Sentinel-1 satellite launched by the Copernicus program of the European Space Agency, using the data type single look complex (SLC) and the polarization method of vertical send and vertical receive (VV). This satellite is equipped with a C-band (5.6 cm wavelength) SAR sensor, the satellite revisit period is 12 days, and the acquisition interval for a few phases of images is 24 days. The imaging mode was interferometric wide-width (IW), with each image having a width of 250 km and resolution of 5 × 20 m. A total of 174 images acquired in the Earth's ascending orbit were used in this study, and the experimental area and coverage of the images are shown in Figure 4a.

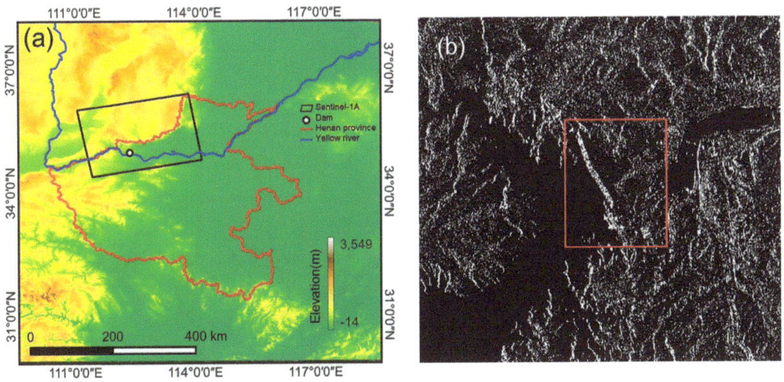


**Figure 4.** (**a**) The schematic diagram of the location of the Sentinel-1 image. The black box in the figure is the coverage area before the image is cropped. (**b**) The SAR image of the dam and the red box is the range of the main body of the dam.

## 3. Methods

### 3.1. InSAR Deformation Model

The InSAR technique uses the phase difference between two satellite passes to obtain a ground digital elevation model [43]. The differential InSAR (D-InSAR) technique introduces an external DEM based on two images to obtain surface deformation information [44]. The application of D-InSAR for regional deformation monitoring is often affected by orbital parameter errors, topographic data errors, phase noise caused by interference loss correlations, phase decoupling errors, and atmospheric delays. To solve this problem,

Ferretti proposed the PS-InSAR technique, which eventually formed the technical theoretical system of MT-InSAR [45–50]. Its classical methodological theory, SBAS-InSAR, can perform phase analysis on coherent targets to obtain time series deformations. Its powerful ability to model time series deformations has been demonstrated in many studies [51–53].

In this study, we used SBAS-InSAR to process Sentinel-1 data and build a time series model of dam deformation, as shown in Figure 5a. After acquiring the raw image data, all the images were corrected using the precision orbit data provided by the ESA. The data were assembled into differential interferometric image pairs according to a time–space baseline of 90 days and 2% of the maximum spatial baseline of all images. The topographic phases were removed using SRTM DEM data at a 90 m resolution which were acquired by the US Space Shuttle. The data were filtered using Goldstein's algorithm [54]. The phase was deconvolved using a minimum-cost flow algorithm [52]. External control points were introduced for trajectory refinement, and the mean displacement rate of the observation points was obtained for terrain correction. The atmospheric phase was estimated and removed in the next step, and the deformation was inverted using the singular value decomposition method to generate a deformation model of the mean displacement rate and time series and finally geocoded to transfer the deformation from the radar coordinate system to the geographic coordinate system.

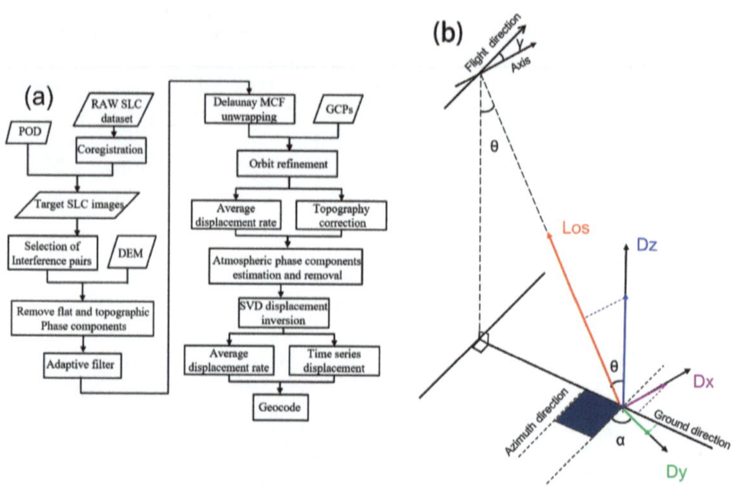


**Figure 5.** (**a**) SBAS-InSAR data processing. (**b**) Projection of the on–site measurement data of the dam deformation at typical point P.

## 3.2. Validation of InSAR Deformation Model Reliability

Synthetic-aperture radar acquires images through side-view imaging, which allows us to obtain the deformation of the observation point in the line of sight (Los) of the satellite. This result expresses the variation in the distance between the observation point and the satellite platform.

The deformation of the observation point in the image is the 3D deformation of the ground point projected onto the image coordinates [55]. Ground truth data can be used to verify the reliability of the InSAR deformation model because the deformation of the ground observation point is usually a joint effect of deformation in multiple directions.

$$\begin{bmatrix} D_{Los} \\ Dazi \end{bmatrix} = \begin{bmatrix} cos\theta & sin\theta sin\alpha & -sin\theta cos\alpha \\ 0 & cos\alpha & sin\alpha \end{bmatrix} \begin{bmatrix} Dz \\ Dx \\ Dy \end{bmatrix} \quad (1)$$

In general, the coordinate system used in geodetic techniques is a geographic coordinate system consisting of north, south, east, and west vertical directions. In this study, we used the measured deformation data of the Xiaolangdi Dam at the on-site point P, and the coordinate system of the data was the local coordinate system of the dam, including the three-dimensional deformation along the dam axis, perpendicular to the dam axis, and in the vertical direction.

We projected the three components of the dam axis, perpendicular to the dam axis, and vertical direction to the satellite's line of sight according to Equation (1), as shown in Figure 5b. Dx represents the dam axis direction component, which is positive to the left bank of the dam; Dy is perpendicular to the dam axis direction component, which is positive downstream of the dam; and Dz represents the vertical direction component, which is positive in the upward direction. The angle between the flight direction of the satellite platform and the dam axis is $\gamma$. $\theta$ is the radar incidence angle, which is the angle between the radar line of sight and the vertical direction. $\alpha$ is the magnitude of the angle of the Dx positive direction clockwise to the satellite flight direction (analogous to the satellite heading angle in the geographic coordinate system).

### 3.3. *InSAR Deformation Prediction Model*

#### 3.3.1. LSTM Neural Network

Deep learning is revolutionizing artificial intelligence around the world [56,57]. An artificial neural network (ANN) is the foundation of deep learning. The artificial neural network model consists of fully connected layers, which can be classified into input, hidden, and output layers. The activation function is at the core of a neural network. The input data are controlled using the activation function. Sigmoid, tanh, and ReLU are the most commonly used activation functions. In this model, the layers are connected by weights. The neural network learns to determine the weights between layers through training. In traditional neural networks, information is transmitted from the input to the output layer. This process can be considered independent. This implies that the output depends only on the current inputs. However, in many realistic tasks, the network output depends not only on the current input but also on its past output.

A recurrent neural network (RNN) takes sequence data as input, iterates in the direction of sequence evolution, and connects all nodes (recurrent units) in a chain fashion. The hidden layer contains a state vector that stores the historical information of all past elements and weights of the entire network. Compared with artificial neural networks, recurrent neural networks can jointly determine the output of the current moment and are more suitable for time series problems, as shown in Figure 6.

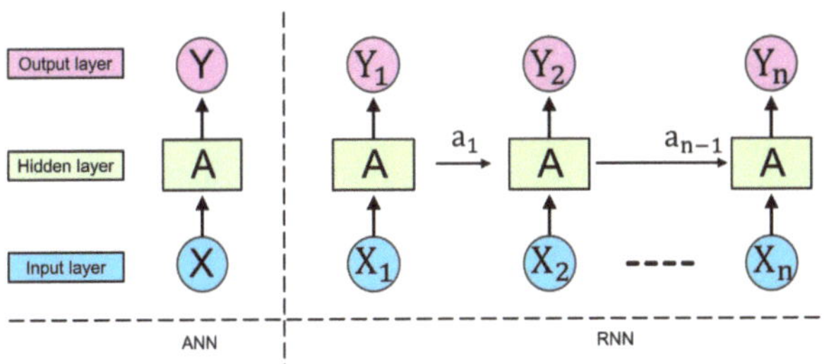


**Figure 6.** The schematic diagram of the model structure of ANN and RNN.

Although recurrent neural networks are used to process time series data, in practice, storing such information for a long time is difficult. The gradient explosion problem occurs in long-term dependency problems. In response, LSTM networks incorporate a special control unit (memory cell) into the recurrent neural network, as shown in Figure 7.

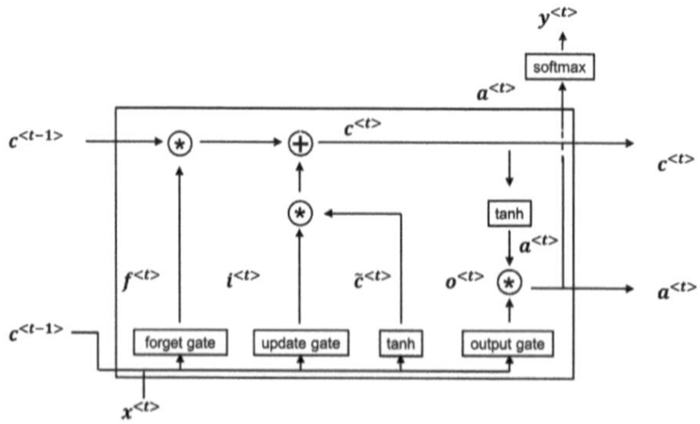


**Figure 7.** The computation process of the LSTM memory unit, the asterisk in the figure indicates the multiplication operation.

The specific computational process is shown in Equations (2a)–(2f), where the memory cell consists of a control gate, update gate, and forget gate. After the data $x^{\langle t\rangle}$ at time $t$, the hidden layer state vector $a^{\langle t-1\rangle}$ and the cell state parameter $c^{\langle t-1\rangle}$ are inputted into the LSTM cell, the candidate value $\tilde{c}^t$ for the cell state parameter update is first generated by the tanh activation function, and then $a^{\langle t-1\rangle}$ and $x^{\langle t\rangle}$ are inputted into the update cell; the update weight $\Gamma_u$, the forget weight $\Gamma_f$, and the output weight $\Gamma_o$ are calculated using the Sigmoid activation function, and the $c^{\langle t\rangle}$ of the next LSTM cell is determined according to $\Gamma_u$ and $\Gamma_f$, while the $a^{\langle t\rangle}$ of the next cell is determined using the output weight $\Gamma_o$. Thus, the LSTM cells are calculated.

$$\tilde{c}^t = tanh\left(W_c\left[a^{\langle t-1\rangle}, x^{\langle t\rangle}\right] + b_c\right) \tag{2a}$$

$$\Gamma_u = \sigma(W_u\left[a^{\langle t-1\rangle}, x^{\langle t\rangle}\right] + b_u \tag{2b}$$

$$\Gamma_f = \sigma(W_f\left[a^{\langle t-1\rangle}, x^{\langle t\rangle}\right] + b_f \tag{2c}$$

$$\Gamma_o = \sigma(W_o\left[a^{\langle t-1\rangle}, x^{\langle t\rangle}\right] + b_o \tag{2d}$$

$$c^{\langle t\rangle} = \Gamma_u * \tilde{c}^t + \Gamma_f * c^{\langle t-1\rangle} \tag{2e}$$

$$a^{\langle t\rangle} = c^{\langle t\rangle} * tanhc^{\langle t\rangle} \tag{2f}$$

### 3.3.2. Construction of the Prediction Model

The structure of the LSTM model used in this study is shown in Figure 8 and was implemented using the TensorFlow and sklearn modules from the Python community. Two LSTM layers were added to the model, each with a size of 50. To prevent model overfitting, a dropout layer was used to discard random nodes in the network at a scale of 0.2. Finally, the final output was transformed into a one-dimensional vector $\widetilde{Y}_i$ using a fully connected layer.

The gradient descent optimization of the loss function was performed using the Adam optimization algorithm, and the mean square error was used to evaluate the accuracy of the model. To train the model, the original data were divided into training and prediction sets, and the k-fold cross-validation method was used for the training set, where k was set to 2. In the training process, the training set was divided into two parts: one part was taken as the validation set, and the other part was taken as the training set. The GridSearchCV function was used to search for the best hyperparameters of the model and obtain the best-weight model. Finally, the test data were fed into the trained model, and the prediction results were output and validated against actual data.

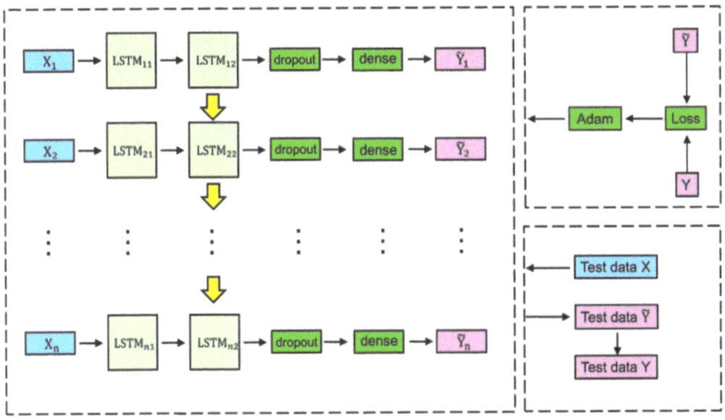


**Figure 8.** The implementation process architecture of the LSTM model used in this study.

## 4. Results and Analysis

### *4.1. InSAR Deformation Results and Validation*

In this study, a time series model of dam deformation was established using Sentinel-1 data from March 2017 to February 2023, and the cumulative deformation of the dam body was obtained, as shown in Figure 9. In the early stages of deformation, the dam tended to be close to the satellite, with a maximum deformation value of 5 mm. In the middle stage of deformation, the center of the dam tended to be far from the satellite, whereas the left and right sides and the bottom of the dam maintained a small tendency to be close to the satellite. In the later stages of deformation, the dam as a whole tended to move away from the satellite, and there were significant differences in the deformation values in different areas, with the largest deformation value of −155 mm occurring in the middle of the top of the dam near the upstream side.

**Figure 9.** The time–series variation in the cumulative deformation of Xiaolangdi Dam from March 2017 to February 2023.

We plotted the cumulative deformation data at different times for points on the upstream slope, top, and downstream slope of the dam on lines, as shown in Figure 10. All three lines tended to have smaller deformation values on both sides, and the central part near the left side had the largest settlement value. The deformation value of the exposed water part of the upstream slope of the dam was slightly larger than that of the top of the dam, and the accumulated deformation values of the downstream slope were significantly different from those of the upstream slope and top of the dam.

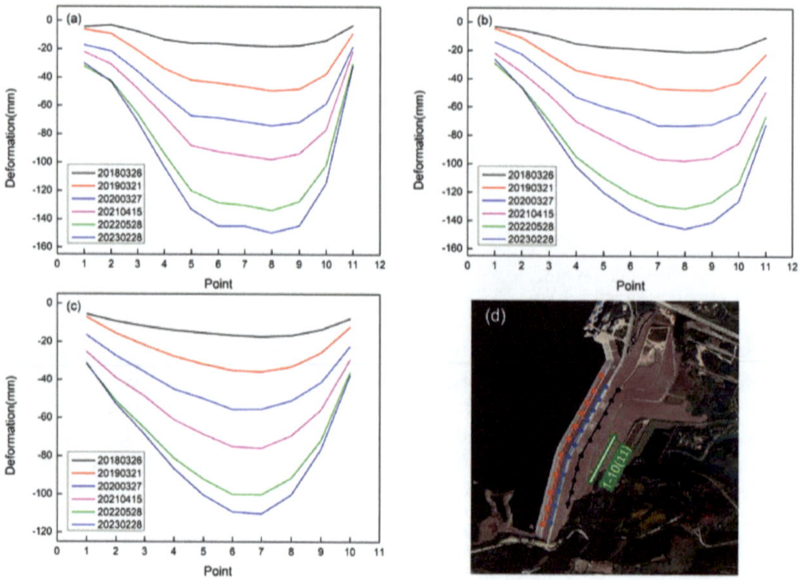


**Figure 10.** (**a**–**c**) Cumulative sink value line segments of the upstream slope of the dam, the top of the dam, and on the downstream slope line, which correspond to the points on the red, blue and black lines of (**d**) respectively. (**d**) Schematic diagram of the location of the three lines; the serial numbers of the points become larger in the direction of the arrow.

The true deformation of the dam obtained via the geodetic method at the typical point P is shown in Figure 11b. The dam field deformation monitoring dataset consists of a total of data in three directions, Dx, Dy, and Dz, acquired from March 2017 to December 2020. We projected the displacement data in three directions upward to the satellite's line of sight according to the formula and compared them with the data obtained via InSAR, as shown in Figure 11c. The accuracy of the accumulated deformation value for P was 95%, and the correlation coefficient R was 0.93. We found that the data distributions of P and the InSAR deformation values were in good agreement.

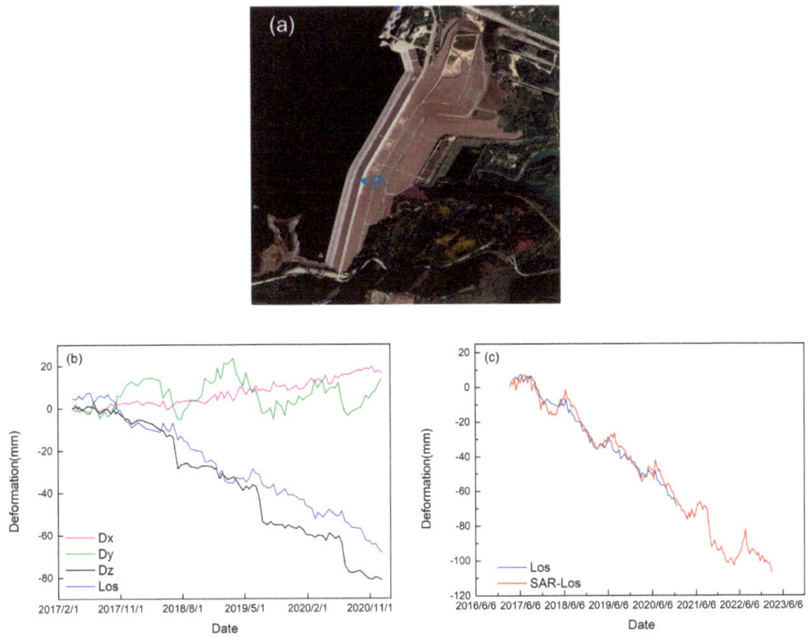


**Figure 11.** (**a**) Location of on-site data point P. (**b**,**c**) Measured data of ground point P; Dx, Dy, and Dz are measured on the ground in each of the three directions, Los is the deformation data Los after the projection of the measured data, and SAR-Los is data from the InSAR deformation model.

## *4.2. Analysis of Deformation and Water Storage Level Data*

Based on the InSAR deformation model data SAR-Los analyzed above, the deformation of the dam was generally decreasing, but there was still a local trend of periodic changes. At the typical point P, the deformation curve had a localized upward trend around June each year, as shown in Figure 12.

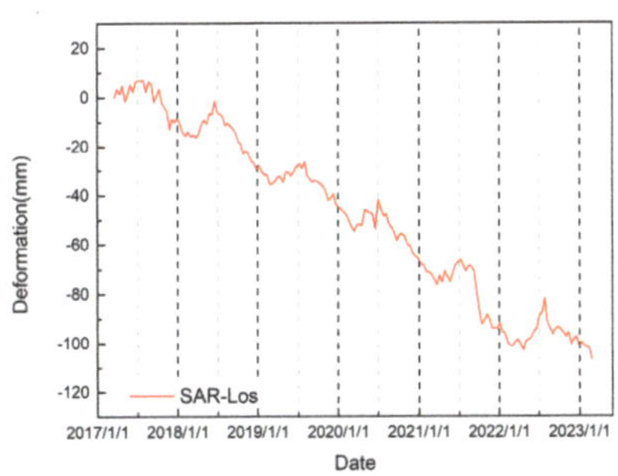


**Figure 12.** The schematic diagram of the periodicity of deformation at point P.

The overall deformation of the dam during 2018–2019 was specifically taken and plotted. As shown in Figure 13, only the center of the dam showed obvious periodic deformation throughout the year, and the overall deformation trends of the center of the dam and point P are the same. But there was basically no deformation on either side or the lower part of the dam. At this point, our conclusion is that the deformation of the dam body of the Xiaolangdi Dam is periodic.

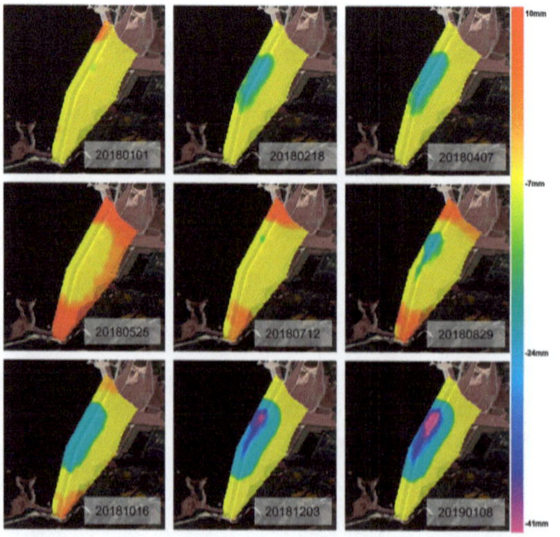


**Figure 13.** The time series variation of the cumulative deformation of Xiaolangdi Dam from 2018 to 2019.

Henan Province has a continental monsoon climate with a large amount of rain in the summer and little rain in the winter. In response to changes in precipitation, reservoir levels must also be adjusted. The water contact surface on the upstream side of the dam is subjected to hydrostatic pressure from the reservoir water. As the reservoir level changes, the hydrostatic pressure also changes. Considering that the Xiaolangdi Dam also exhibits a cyclic deformation trend, we speculate that hydrostatic pressure is the main driving factor for deformation.

We compared the InSAR deformation data of the dam at the typical point P with the water level data, as shown in Figure 14. The Pearson correlation coefficient R between the accumulated deformation values of the InSAR model and the storage level data was 0.81. In order to express the relationship between the two more clearly, we selected the detailed comparison data of the storage level and deformation between 2018 and 2019 for analysis.

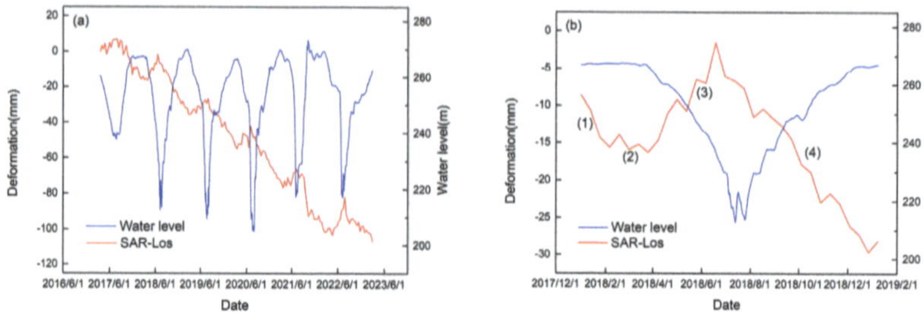


**Figure 14.** (**a**) Comparison of deformation data and water storage level data at point P; (**b**) Stage diagram of deformation data and water storage level data in 2018–2019, (1)–(4) are the numbering of the different stages based on the trend of the line.

Over one year, the deformation curve can be roughly divided into four stages according to the slope of the curve, as shown in Figure 14b, and we performed a statistical analysis on the water storage level and deformation information in different stages, as shown in Table 1. By analyzing the relationship between the two, we concluded that around the field measuring point, the deformation value decreases when the water storage level increases and increases when the water storage level decreases around the typical point P.

**Table 1.** The characteristics of the water storage level and dam deformation changes at different times.

| Stage | 1 | 2 | 3 | 4 |
|---|---|---|---|---|
| Date | 1 January 2018–6 February 2018 | 6 February 2018–26 March 2018 | 26 March 2018–18 June 2018 | 18 June 2018–8 January 2019 |
| Average daily level | 267.2 m | 267.2 m | 253.5 m | 246.5 m |
| Average daily level change | +0.01 m | −0.02 m | −0.35 m | +0.15 m |
| Cumulative deformation value | −7 mm | −0.7 mm | +14.8 mm | −26.6 mm |
| Daily deformation rate | −0.19 mm | −0.01 mm | +0.18 mm | −0.13 mm |

### *4.3. Prediction of Deformation Based on Reservoir Water Level Data*

The Xiaolangdi Dam is subject to periodic deformation owing to changes in the upstream reservoir level. The storage level and deformation showed a strong correlation.

In this study, data from in situ measurements were only available at the typical point P, and the monitoring period was short and irregular. In contrast, the InSAR deformation model has a large amount of data and a higher accuracy, which was verified by the data at the typical point P.

The time span of the InSAR data used in the time series deformation model was six years, with a total of 173 periods of deformation data and a total of 2386 points for each period of InSAR deformation data. After all data were interpolated, a total of 181 data periods were obtained, with an interval of 12 d between two data periods. Each InSAR deformation period corresponds to twelve days of daily reservoir level data.

The InSAR-LSTM deformation prediction model was built by taking the daily reservoir level data as input and the InSAR model deformation data as output. The data were divided into a training set and prediction set according to the time series, and the division results are shown in Table 2. The hyperparameter steps were applied in the model to combine the data, and after some debugging, the best step parameter was determined to be 2, which means that every two sets of two periods of connected data were combined, and the final ratio of the training set and the prediction set obtained was 164:13.

**Table 2.** Distribution of training and prediction sets in the total data and their time spans.

| | Number | Start Date | End Date |
|---|---|---|---|
| Training set | 166/164 | 31 March 2017 | 25 September 2022 |
| Prediction set | 15/13 | 7 October 2022 | 28 February 2023 |

The LSTM model produced thirteen sets of predictions; Figure 15a shows the first period of data for the predicted results, which is the cumulative deformation data for the dam on 7 October 2022. Figure 15c shows the errors of the InSAR deformation data y and the prediction data $y_{pred}$, which are shown in Figure 15b. The error calculation index is MAE, which is calculated as in Equation (4); the relative error is still concentrated at 10 percent or less, and only some of the points of the absolute error reach 24 mm.

$$R = \frac{\Sigma(x-\overline{x})(x=y-\overline{y})}{\sqrt{\Sigma(x-\overline{x})^2(y-\overline{y})^2}} \quad (3)$$

$$MAE = \frac{1}{n}\sum_{i=1}^{n}\left|y^i - y^i{}_{pred}\right| \quad (4)$$

$$MSE = \frac{1}{n}\sum_{i=1}^{n}\left(y^i - y^i{}_{pred}\right)^2 \quad (5)$$

$$RMSE = \sqrt{\frac{1}{n}\sum_{i=1}^{n}\left(y^i - y^i{}_{pred}\right)^2} \quad (6)$$

At the typical point P, the comparison between the InSAR deformation data and the prediction data is shown in Figure 15d. There are thirteen periods of data for point P. The final cumulative deformation values for the InSAR deformation data and the predicted deformation data were −106.8 mm and −103 mm, respectively, and the relative accuracies were 96.4%. We used MAE, MSE, RMSE, and R as indicators for evaluating the overall prediction accuracy, which are calculated as shown in Equations (3)–(6). The calculation results are shown in Table 3.

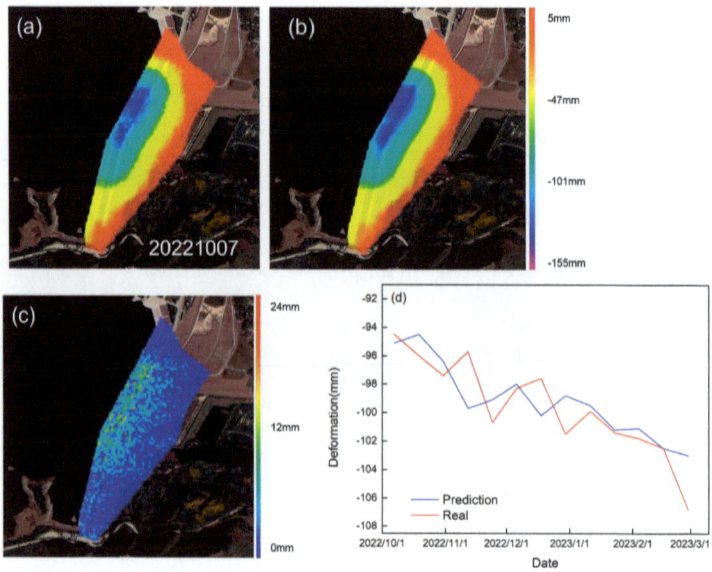


**Figure 15.** (**a**) Prediction results deformation data. (**b**) InSAR model deformation data. (**c**) Error distribution of the MAE in the prediction results. The results of the remaining twelve forecasts are presented in the Supplementary document. (**d**) Comparison of deformation data and predicted results at point P.

**Table 3.** Accuracy evaluation for four deformation prediction models.

| | MAE | MSE | RMSE | R |
|---|---|---|---|---|
| ANN | 3.67 | 19.51 | 4.41 | −0.56 |
| RNN | 4.56 | 25.93 | 5.09 | 0.72 |
| LSTM | 1.49 | 3.95 | 1.98 | 0.80 |
| LSTM-Tem | 1.37 | 3.45 | 1.85 | 0.83 |

### 4.4. Multimodel Comparison and Parameter Optimization

In Figure 16a, the prediction data of the ANN, RNN, and LSTM models are compared with the InSAR deformation data, and the accuracy evaluation indexes of each model are shown in Table 3. It can be seen that the LSTM model had the highest prediction accuracy and the ANN model had the lowest prediction accuracy, which proves that the LSTM model is more suitable for deformation prediction in this study.

At present, researchers studying dam deformation commonly use the rheological deformation model, which is a real physical model. This approach requires a large number of dam-related parameters and a high level of professionalism on the part of the researcher [58].

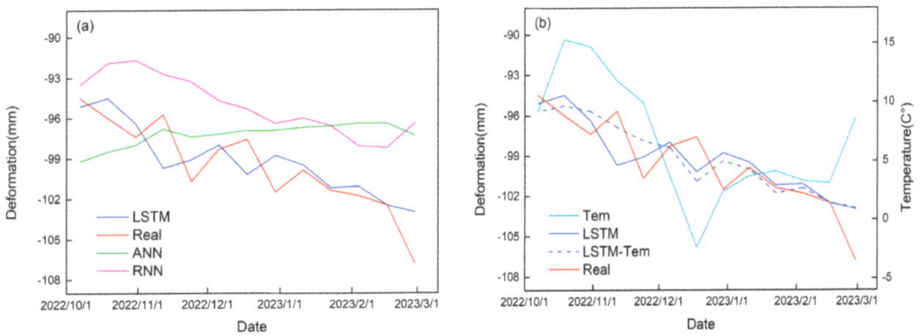


**Figure 16.** (**a**) Comparison of the prediction data of the three models ANN, RNN and LSTM at typical point P. (**b**) Comparison of the prediction data before and after the addition of the temperature parameter to the LSTM model.

The LSTM model is a mathematical model that can be modeled from a numerical point of view, but it does not take into account the real physical meaning of the numerical values in the calculation. In addition to hydrostatic pressure, environmental parameters such as temperature also have an effect on the deformation of the dam. We optimized the LSTM deformation prediction model using daily average temperature data.

Figure 16b shows the predicted data using the model with the added parameters compared to the data without the added parameters, and a comparison of the accuracy of the two sets of predicted and deformed data is shown in Table 3. From the accuracy index in the table, it can be seen that adding the temperature environment parameter to the LSTM deformation prediction model can make the prediction more closely match the deformation data.

## 5. Discussion

The main body of the dam sinks under the influence of its own gravity. In the case of a rockfill dam, which is constructed of rocks and sediment, the water stored in the reservoir percolates from upstream to downstream, creating a seepage field. The seepage field exerts uplift pressure on the bottom of the dam, which is equivalent to reducing the gravity of the dam. In addition, the reservoir impoundment exerts hydrostatic pressure on the contact surface between the dam and the water, and the dam will slowly move downstream, with a change in hydrostatic pressure and a change in the trend of movement.

The relationship between on-site measurement displacement and storage level in the three directions of the typical point P is shown in Figure 17a. The amount of displacement of point P to the left and right along the axis of the dam was relatively small and showed slow growth. In the downstream direction of the dam, the magnitude of the displacement was basically synchronized with the rise and fall of the storage level, but there was still a delay of about 20 days. In the vertical direction of the dam, an accelerated subsidence of point P occurred each year after the water level fell since the uplift pressure was reduced and slowly continued as the water level began to rise to its highest point. Compared with the vertical settlement, the displacement along the flow direction of the dam contributed to a larger part in the projection process. Therefore, in the process of moving away from the satellite platform, the dam shows a small rebound trend with a decrease in the storage water level, but the overall trend is away from the satellite platform.

From the perspective of this study's findings, if we want to reduce the deformation of the dam, we should try to avoid storing the reservoir at the highest level for a long period of time in order to mitigate the movement of the dam downstream, restore the water after the reservoir storage level decreases, use the uplift pressure to mitigate the effect of gravity on the settlement of the dam, and avoid the prolonged operation of the reservoir at a low water level.

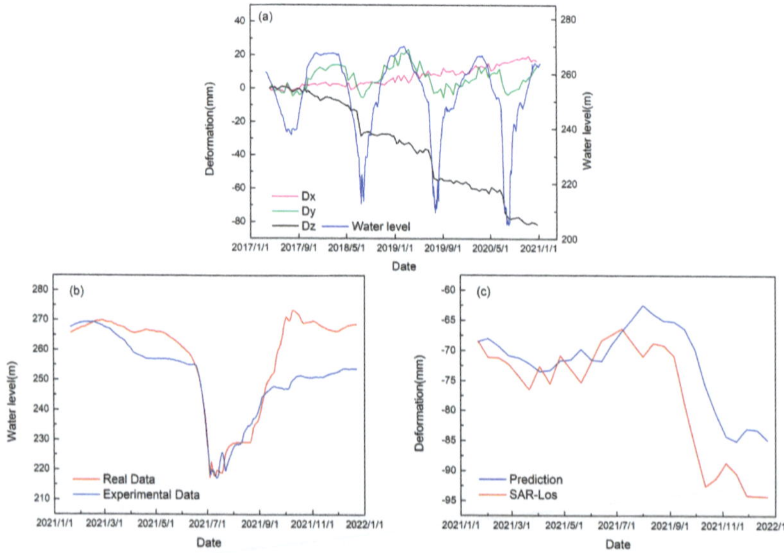


**Figure 17.** (**a**) Relationship between on-site deformation data and storage level data at typical point P. (**b**,**c**) Comparison of the point P deformation values predicted using the simulated storage level experimental data with the InSAR model deformation values in 2021.

The annual cumulative deformation value of point P in 2021 reached −25.9 mm, which is the maximum value of annual deformation in the whole monitored cycle. We made adjustments to the actual storage level in 2021 as described above, and Figure 15b shows a comparison of the simulated storage with the real storage level data after the optimization was performed. We used the optimized storage level experimental data as input into the deformation prediction model, and the results are shown in Figure 17c. At each of the 29 points shown in the figure, the average deformation weakening value was 4.9 mm. The final cumulative annual deformation value was −16.4 mm, which is an effective deformation reduction of about 37%.

The Xiaolangdi Water Conservancy Hub is a large-scale water conservancy facility, and effective reservoir storage management can extend the service life of the dam, and any changes to the reservoir level management program must be carefully considered before they are put into practice. In general, reservoir managers should follow a multifactorial approach to determine the reservoir's storage level scheduling program, such as using actual river flow and weather data. When the weather is bad and the reservoir needs to be kept in operation at an extreme storage level, the model can also be used in advance to predict possible pitfalls and determine countermeasures.

## 6. Conclusions

The Xiaolangdi Dam was constructed more than 20 years ago, and the main body of the dam underwent deformation during 2017–2023, which was investigated in this study. A time series deformation model of the main body of the Xiaolangdi Dam was established using InSAR technology, which can accurately reflect the overall deformation information of the dam after verification using the measured data. By combining the ground truth data, the InSAR time series deformation model, and the LSTM deformation prediction model, we can conclude the following:

1. The InSAR deformation model shows that there is a gradual weakening in the deformation trend of the dam from the center to the sides and from the top to the bottom. Throughout the 6-year deformation cycle, although there were differences in the

deformation trends in different parts of the dam, each region was excessively smooth. The 6-year cumulative deformation in the middle part of the dam near the upstream reached -155 mm, which is within the safe range for large rockfill dams.
2. The Xiaolangdi Dam continuously deforms. The satellite platform can continuously and periodically acquire InSAR image data, which helps monitor the overall deformation of the dam over a long period of time and allows more deformation information to be obtained. Theoretically, the combination of InSAR technology and the LSTM model can predict the effects of different storage level planning schemes on the dam and can then adjust storage level planning schemes in a targeted manner, attenuating dam deformation and preventing the risk of possible larger deformations.
3. Owing to the inherent limitations of the satellite platform, ground-based measurement data are also required to verify the reliability of the deformation and prediction models. In the future, the launch of satellites with shorter revisit periods and higher resolutions could enable better monitoring of surface deformation. The specific mechanism by which hydrostatic pressure affects the structural stability of dams has not been studied in depth in this work, and this could be the subject of future research.

**Supplementary Materials:** The following supporting information can be downloaded at: https://www.mdpi.com/article/10.3390/w15193384/s1. Figure S1: Total dam deformation prediction results.

**Author Contributions:** Conceptualization, Z.F., Y.P. and R.H.; methodology, Z.F.; software, Z.F.; validation, Z.F. and H.Y.; formal analysis, R.H. and Y.P.; investigation, H.Y. and Z.H; resources, R.H.; data curation, Z.F.; writing—original draft preparation, Z.F.; writing—review and editing, R.H., Z.H., Y.P. and H.Y.; visualization, Z.F.; supervision, Z.H. and Y.P. All authors have read and agreed to the published version of the manuscript.

**Funding:** This work was supported by the State Key Project of the National Natural Science Foundation of China, Key Projects of Joint Funds for Regional Innovation and Development (grant number U22A20566); the Henan Provincial Higher Education Key Research Funding Project (grant number 18B420003); and the Henan University of Science and Technology Basic Research Business Expenses Specially Funded Project (grant number NSFRF170909).

**Data Availability Statement:** Santinel-1 image data and reservoir storage level data acquisition methods are mentioned in the Acknoledgments.

**Acknowledgments:** We sincerely thank the ESA for providing the Sentinel-1 data (https://search.asf.alaska.edu [accessed on 1 September 2023]), Guangqian Hu and Huijun Yang of the Xiaolangdi Engineering Consulting Co., Ltd., for compiling the measured data on the deformation of the Xiaolangdi Dam, and the Yellow River Network of the Yellow River Conservancy Commission of the Ministry of Water Resources of China for providing information on the storage level of the Xiaolangdi Reservoir (www.yrcc.gov.cn [accessed on 1 September 2023]).

**Conflicts of Interest:** The authors declare no conflict of interest.

## References

1. Ilyushin, Y.V.; Kapostey, E.I. Developing a Comprehensive Mathematical Model for Aluminium Production in a Soderberg Electrolyser. *Energies* **2023**, *16*, 6313. [CrossRef]
2. Pershin, I.M.; Papush, E.G.; Kukharova, T.V.; Utkin, V.A. Modeling of Distributed Control System for Network of Mineral Water Wells. *Water* **2023**, *15*, 2289. [CrossRef]
3. Wang, Q.Q.; Huang, Q.H.; He, N.; He, B.; Wang, Z.C.; Wang, Y.A. Displacement monitoring of upper Atbara dam based on time series InSAR. *Surv. Rev.* **2020**, *52*, 485–496. [CrossRef]
4. Milillo, P.; Perissin, D.; Salzer, J.T.; Lundgren, P.; Lacava, G.; Milillo, G.; Serio, C. Monitoring dam structural health from space: Insights from novel InSAR techniques and multi-parametric modeling applied to the Pertusillo dam Basilicata, Italy. *Int. J. Appl. Earth Obs. Geoinf.* **2016**, *52*, 221–229. [CrossRef]
5. Rotta, L.H.S.; Alcântara, E.; Park, E.; Negri, R.G.; Lin, Y.N.; Bernardo, N.; Mendes, T.S.G.; Filho, C.R.S. The 2019 Brumadinho tailings dam collapse: Possible cause and impacts of the worst human and environmental disaster in Brazil. *Int. J. Appl. Earth Obs. Geoinf.* **2020**, *90*, 102119.
6. Gikas, V.; Sakellariou, M. Settlement analysis of the Mornos earth dam (Greece): Evidence from numerical modeling and geodetic monitoring. *Eng. Struct.* **2008**, *30*, 3074–3081. [CrossRef]

7. Xi, R.; Liang, Y.; Chen, Q.; Jiang, W.; Chen, Y.; Liu, S. Analysis of Annual Deformation Characteristics of Xilongchi Dam Using Historical GPS Observations. *Remote Sens.* **2022**, *14*, 4018. [CrossRef]
8. Chen, Z.; Yin, Y.; Yu, J.; Cheng, X.; Zhang, D.; Li, Q. Internal deformation monitoring for earth-rockfill dam via high-precision flexible pipeline measurements. *Autom. Constr.* **2022**, *136*, 104177. [CrossRef]
9. Yao, F.H.; Guan, S.H.; Yang, H.; Chen, Y.; Qiu, H.F.; Ma, G.; Liu, Q.W. Long-term deformation analysis of Shuibuya concrete face rockfill dam based on response surface method and improved genetic algorithm. *Water Sci. Eng.* **2019**, *12*, 196–204. [CrossRef]
10. Ma, C.; Yang, J.; Cheng, L.; Ran, L. Adaptive parameter inversion analysis method of rockfill dam based on harmony search algorithm and mixed multi-output relevance vector machine. *Eng. Comput.* **2020**, *37*, 2229–2249. [CrossRef]
11. Xue, F.; Lv, X.; Dou, F.; Yun, Y. A Review of Time-Series Interferometric SAR Techniques: A Tutorial for Surface Deformation Analysis. *IEEE Geosci. Remote Sens. Mag.* **2020**, *8*, 22–42. [CrossRef]
12. Li, B.; Li, Y.; Jiang, W.; Su, Z.; Shen, W. Conjugate ruptures and seismotectonic implications of the 2019 Mindanao earthquake sequence inferred from Sentinel-1 InSAR data. *Int. J. Appl. Earth Obs. Geoinf.* **2020**, *90*, 102127. [CrossRef]
13. Yang, Y.; Chen, Q.; Xu, Q.; Zhang, Y.; Yong, Q.; Liu, G. Coseismic surface deformation of the 2014 Napa earthquake mapped by Sentinel-1A SAR and accuracy assessment with COSMO-SkyMed and GPS data as cross validation. *Int. J. Digit. Earth* **2017**, *10*, 1197–1213. [CrossRef]
14. Noviello, C.; Verde, S.; Zamparelli, V.; Fornaro, G.; Pauciullo, A.; Reale, D.; Nicodemo, G.; Ferlisi, S.; Gulla, G.; Peduto, D. Monitoring Buildings at Landslide Risk With SAR: A Methodology Based on the Use of Multipass Interferometric Data. *IEEE Geosci. Remote Sens. Mag.* **2020**, *8*, 91–119. [CrossRef]
15. Rodríguez, Á.B.; Balestriero, R.; De Angelis, S.; Benítez, M.C.; Zuccarello, L.; Baraniuk, R.; Ibanez, J.M.; de Hoop, M.V. Recurrent Scattering Network Detects Metastable Behavior in Polyphonic Seismo-Volcanic Signals for Volcano Eruption Forecasting. *IEEE Trans. Geosci. Remote Sens.* **2022**, *60*, 5909123.
16. An, B.; Jiang, Y.; Wang, C.; Shen, P.; Song, T.; Hu, C.; Liu, K. Ground infrastructure monitoring in coastal areas using time-series inSAR technology: The case study of Pudong International Airport, Shanghai. *Int. J. Digit. Earth* **2023**, *16*, 2171144. [CrossRef]
17. Shi, X.; Zhu, T.; Tang, W.; Jiang, M.; Jiang, H.; Yang, C.; Zhan, W.; Ming, Z.; Zhang, S. Inferring decelerated land subsidence and groundwater storage dynamics in Tianjin–Langfang using Sentinel-1 InSAR. *Int. J. Digit. Earth* **2022**, *15*, 1526–1546. [CrossRef]
18. Pawłuszek-Filipiak, K.; Wielgocka, N.; Tondaś, D.; Borkowski, A. Monitoring nonlinear and fast deformation caused by underground mining exploitation using multi-temporal Sentinel-1 radar interferometry and corner reflectors: Application, validation and processing obstacles. *Int. J. Digit. Earth* **2023**, *16*, 251–271. [CrossRef]
19. Yang, Z.; Xu, B.; Li, Z.; Wu, L.; Zhu, J. Prediction of Mining-Induced Kinematic 3-D Displacements from InSAR Using a Weibull Model and a Kalman Filter. *IEEE Trans. Geosci. Remote Sens.* **2022**, *60*, 4500912. [CrossRef]
20. Liu, L.; Yu, J.; Chen, B.; Wang, Y. Urban subsidence monitoring by SBAS-InSAR technique with multi-platform SAR images: A case study of Beijing Plain, China. *Eur. J. Remote Sens.* **2020**, *53* (Suppl. 1), 141–153. [CrossRef]
21. Xing, X.; Zhu, Y.; Xu, W.; Peng, W.; Yuan, Z. Measuring Subsidence Over Soft Clay Highways Using a Novel Time-Series InSAR Deformation Model with an Emphasis on Rheological Properties and Environmental Factors (NREM). *IEEE Trans. Geosci. Remote Sens.* **2022**, *60*, 4601319. [CrossRef]
22. Xing, X.; Huang, L.; He, Z.; Zhang, T.; Zhu, Y. Health Observation of the Capital Airport South Expressway Based on Improved MT-InSAR Technology. *IEEE J. Miniaturization Air Space Syst.* **2023**, *4*, 232–241. [CrossRef]
23. Tang, W.; Motagh, M.; Zhan, W. Monitoring active open-pit mine stability in the Rhenish coalfields of Germany using a coherence-based SBAS method. *Int. J. Appl. Earth Obs. Geoinf.* **2020**, *93*, 102217. [CrossRef]
24. Fan, H.; Li, T.; Gao, Y.; Deng, K.; Wu, H. Characteristics inversion of underground goaf based on InSAR techniques and PIM. *Int. J. Appl. Earth Obs. Geoinf.* **2021**, *103*, 102526. [CrossRef]
25. Wang, Q.; Fan, J.; Zhou, W.; Tong, L.; Guo, Z.; Liu, G.; Yuan, W.; Sousa, J.J.; Perski, Z. 3D Surface velocity retrieval of mountain glacier using an offset tracking technique applied to ascending and descending SAR constellation data: A case study of the Yiga Glacier. *Int. J. Digit. Earth* **2019**, *12*, 614–624. [CrossRef]
26. Li, Z.; Wu, Q. Capturing the crack process of the Antarctic A74 iceberg with Sentinel-1 based offset tracking and radar interferometry techniques. *Int. J. Digit. Earth* **2022**, *15*, 397–415. [CrossRef]
27. Zhang, Z.; Lin, H.; Wang, M.; Liu, X.; Chen, Q.; Wang, C.; Zhang, H. A Review of Satellite Synthetic Aperture Radar Interferometry Applications in Permafrost Regions: Current status, challenges, and trends. *IEEE Geosci. Remote Sens. Mag.* **2022**, *10*, 93–114. [CrossRef]
28. Zhou, W.; Li, S.; Zhou, Z.; Chang, X. InSAR Observation and Numerical Modeling of the Earth-Dam Displacement of Shuibuya Dam (China). *Remote Sens.* **2016**, *8*, 877. [CrossRef]
29. Zhou, W.; Li, S.; Zhou, Z.; Chang, X. Remote Sensing of Deformation of a High Concrete-Faced Rockfill Dam Using InSAR: A Study of the Shuibuya Dam, China. *Remote Sens.* **2016**, *8*, 255. [CrossRef]
30. Ruiz-Armenteros, A.M.; Marchamalo-Sacrsitán, M.; Bakoň, M.; Lamas-Fernández, F.; Delgado, J.M.; Sánchez-Ballesteros, V.; Papco, J.; González-Rodrigo, B.; Lazecky, M.; Perissin, D.; et al. Monitoring of an embankment dam in southern Spain based on Sentinel-1 Time-series InSAR. *Procedia Comput. Sci.* **2021**, *181*, 353–359. [CrossRef]
31. Biondi, F.; Addabbo, P.; Clemente, C.; Ullo, S.L.; Orlando, D. Monitoring of Critical Infrastructures by Micromotion Estimation: The Mosul Dam Destabilization. *IEEE J. Sel. Top. Appl. Earth Obs. Remote Sens.* **2020**, *13*, 6337–6351. [CrossRef]

Xiao, R.; Jiang, M.; Li, Z.; He, X. New insights into the 2020 Sardoba dam failure in Uzbekistan from Earth observation. *Int. J. Appl. Earth Obs. Geoinf.* **2022**, *107*, 102705. [CrossRef]
Bayik, C.; Abdikan, S.; Arıkan, M. Long term displacement observation of the Atatürk Dam, Turkey by multi-temporal InSAR analysis. *Acta Astronaut.* **2021**, *189*, 483–491. [CrossRef]
Hochreiter, S.; Schmidhuber, J. Long Short-Term Memory. *Neural Comput.* **1997**, *9*, 1735–1780. [CrossRef] [PubMed]
Senanayake, S.; Pradhan, B.; Alamri, A.; Park, H.-J. A new application of deep neural network (LSTM) and RUSLE models in soil erosion prediction. *Sci. Total Environ.* **2022**, *845*, 157220. [CrossRef] [PubMed]
Dikshit, A.; Pradhan, B.; Alamri, A.M. Pathways and challenges of the application of artificial intelligence to geohazards modelling. *Gondwana Res.* **2021**, *100*, 290–301. [CrossRef]
Guo, Y.; Yu, X.; Xu, Y.-P.; Chen, H.; Gu, H.; Xie, J. AI-based techniques for multi-step streamflow forecasts: Application for multi-objective reservoir operation optimization and performance assessment. *Hydrol. Earth Syst. Sci.* **2021**, *25*, 5951–5979. [CrossRef]
Huang, B.; Kang, F.; Li, J.; Wang, F. Displacement prediction model for high arch dams using long short-term memory based encoder-decoder with dual-stage attention considering measured dam temperature. *Eng. Struct.* **2023**, *280*, 115686. [CrossRef]
Jena, R.; Naik, S.P.; Pradhan, B.; Beydoun, G.; Park, H.-J.; Alamri, A. Earthquake vulnerability assessment for the Indian subcontinent using the Long Short-Term Memory model (LSTM). *Int. J. Disaster Risk Reduct.* **2021**, *66*, 102642. [CrossRef]
Li, H.; Xu, Q.; He, Y.; Fan, X.; Yang, H.; Li, S. Temporal detection of sharp landslide deformation with ensemble-based LSTM-RNNs and Hurst exponent. *Geomat. Nat. Hazards Risk* **2021**, *12*, 3089–3113. [CrossRef]
Chen, C.; Zhang, Q.; Kashani, M.H.; Jun, C.; Bateni, S.M.; Band, S.S.; Dash, S.S.; Chau, K.W. Forecast of rainfall distribution based on fixed sliding window long short-term memory. *Eng. Appl. Comput. Fluid Mech.* **2022**, *16*, 248–261. [CrossRef]
Radman, A.; Akhoondzadeh, M.; Hosseiny, B. Integrating InSAR and deep-learning for modeling and predicting subsidence over the adjacent area of Lake Urmia, Iran. *GISci. Remote Sens.* **2021**, *58*, 1413–1433. [CrossRef]
Zebker, H.; Werner, C.; Rosen, P.; Hensley, S. Accuracy of topographic maps derived from ERS-1 interferometric radar. *IEEE Trans. Geosci. Remote Sens.* **1994**, *32*, 823–836. [CrossRef]
Zhang, Z.; Zeng, Q.; Jiao, J. Application of D-InSAR Technology on Risk Assessment of Mining Area. In Proceedings of the IGARSS 2019–2019 IEEE International Geoscience and Remote Sensing Symposium, Yokohama, Japan, 28 July–2 August 2019; pp. 9695–9698.
Lanari, R.; Mora, O.; Manunta, M.; Mallorqui, J.; Berardino, P.; Sansosti, E. A small-baseline approach for investigating deformations on full-resolution differential SAR interferograms. *IEEE Trans. Geosci. Remote Sens.* **2004**, *42*, 1377–1386. [CrossRef]
Hooper, A.; Prata, F.; Sigmundsson, F. Remote Sensing of Volcanic Hazards and Their Precursors. *Proc. IEEE* **2012**, *100*, 2908–2930. [CrossRef]
Ferretti, A.; Prati, C.; Rocca, F. Permanent scatterers in SAR interferometry. *IEEE Trans. Geosci. Remote Sens.* **2001**, *39*, 8–20. [CrossRef]
Ferretti, A.; Prati, C.; Rocca, F. Nonlinear subsidence rate estimation using permanent scatterers in differential SAR interferometry. *IEEE Trans. Geosci. Remote Sens.* **2000**, *38*, 2202–2212. [CrossRef]
Berardino, P.; Fornaro, G.; Lanari, R.; Sansosti, E. A new algorithm for surface deformation monitoring based on small baseline differential SAR interferograms. *IEEE Trans. Geosci. Remote Sens.* **2002**, *40*, 2375–2383. [CrossRef]
Liu, G.; Buckley, S.M.; Ding, X.; Chen, Q.; Luo, X. Estimating Spatiotemporal Ground Deformation with Improved Permanent-Scatterer Radar Interferometry. *IEEE Trans. Geosci. Remote Sens.* **2009**, *47*, 2762–2772. [CrossRef]
Tao, Q.; Wang, F.; Guo, Z.; Hu, L.; Yang, C.; Liu, T. Accuracy verification and evaluation of small baseline subset (SBAS) interferometric synthetic aperture radar (InSAR) for monitoring mining subsidence. *Eur. J. Remote Sens.* **2021**, *54*, 642–663. [CrossRef]
Pepe, A.; Euillades, L.D.; Manunta, M.; Lanari, R. New Advances of the Extended Minimum Cost Flow Phase Unwrapping Algorithm for SBAS-DInSAR Analysis at Full Spatial Resolution. *IEEE Trans. Geosci. Remote Sens.* **2011**, *49*, 4062–4079. [CrossRef]
Chen, F.; Zhou, W.; Tang, Y.; Li, R.; Lin, H.; Balz, T.; Luo, J.; Shi, P.; Zhu, M.; Fang, C. Remote sensing-based deformation monitoring of pagodas at the Bagan cultural heritage site, Myanmar. *Int. J. Digit. Earth* **2022**, *15*, 770–788. [CrossRef]
Song, R.; Guo, H.; Liu, G.; Perski, Z.; Yue, H.; Han, C.; Fan, J. Improved Goldstein SAR Interferogram Filter Based on Adaptive-Neighborhood Technique. *IEEE Geosci. Remote Sens. Lett.* **2015**, *12*, 140–144. [CrossRef]
Hu, J.; Li, Z.-W.; Sun, Q.; Zhu, J.-J.; Ding, X.-L. Three-Dimensional Surface Displacements from InSAR and GPS Measurements With Variance Component Estimation. *IEEE Geosci. Remote Sens. Lett.* **2012**, *9*, 754–758.
LeCun, Y.; Bengio, Y.; Hinton, G. Deep learning. *Nature* **2015**, *521*, 436–444. [CrossRef]
Chiang, Y.-M.; Hao, R.-N.; Xu, Y.-P.; Liu, L. Multi-source rainfall merging and reservoir inflow forecasting by ensemble technique and artificial intelligence. *J. Hydrol. Reg. Stud.* **2022**, *44*, 101204. [CrossRef]
Qiu, Z.; Cao, T.; Li, Y.; Wang, J.; Chen, Y. Rheological Behavior and Modeling of a Crushed Sandstone-Mudstone Particle Mixture. *Processes* **2018**, *6*, 192. [CrossRef]

MDPI AG
Grosspeteranlage 5
4052 Basel
Switzerland
Tel.: +41 61 683 77 34

*Water* Editorial Office
E-mail: water@mdpi.com
www.mdpi.com/journal/water

www.ingramcontent.com/pod-product-compliance
Lightning Source LLC
Chambersburg PA
CBHW041348210926
51850CB00013B/3024